Enchanted Calvinism

Rochester Studies in African History and the Diaspora

Toyin Falola, Series Editor
The Jacob and Frances Sanger Mossiker Chair in the
Humanities and University Distinguished Teaching Professor
University of Texas at Austin

(ISSN: 1092-5228)

Ira Aldridge: The Early Years, 1807–1833
Bernth Lindfors

Ira Aldridge: The Vagabond Years, 1833–1852
Bernth Lindfors

African Police and Soldiers in Colonial Zimbabwe, 1923–80
Timothy Stapleton

Globalization and Sustainable Development in Africa
Edited by Bessie House-Soremekun and Toyin Falola

The Fante and the Transatlantic Slave Trade
Rebecca Shumway

Western Frontiers of African Art
Moyo Okediji

Women and Slavery in Nineteenth-Century Colonial Cuba
Sarah L. Franklin

*Ethnicity in Zimbabwe: Transformations
in Kalanga and Ndebele Societies, 1860–1990*
Enocent Msindo

Edward Wilmot Blyden and the Racial Nationalist Imagination
Teshale Tibebu

South Africa and the World Economy: Remaking Race, State, and Region
William G. Martin

A complete list of titles in the Rochester Studies in African History and the Diaspora series may be found on our website, www.urpress.com.

Enchanted Calvinism

Labor Migration, Afflicting Spirits, and Christian Therapy in the Presbyterian Church of Ghana

Adam Mohr

UNIVERSITY OF ROCHESTER PRESS

Copyright © 2013 by Adam Mohr

All rights reserved. Except as permitted under current legislation, no part of this work may be photocopied, stored in a retrieval system, published, performed in public, adapted, broadcast, transmitted, recorded, or reproduced in any form or by any means, without the prior permission of the copyright owner.

First published 2013

University of Rochester Press
668 Mt. Hope Avenue, Rochester, NY 14620, USA
www.urpress.com
and Boydell & Brewer Limited
PO Box 9, Woodbridge, Suffolk IP12 3DF, UK
www.boydellandbrewer.com

ISBN-13: 978-1-58046-462-8
ISSN: 1092-5228

Library of Congress Cataloging-in-Publication Data

Mohr, Adam
  Enchanted calvinism: labor migration, afflicting spirits, and Christian therapy in the Presbyterian Church of Ghana / Adam Mohr.
      p. cm. — (Rochester studies in African history and the diaspora, ISSN 1092-5228 ; v 58)
   Includes bibliographical references and index.
   ISBN 978-1-58046-462-8 (hardcover: alk. paper) 1. Presbyterian Church of Ghana. 2. Spiritual healing—Ghana. 3. Spiritual healing—Christianity. 4. Healing—Ghana—Religious aspects. I. Title.
   LC-classification not assigned
   203.109667—dc23

2013023532

A catalogue record for this title is available from the British Library.

This publication is printed on acid-free paper.
Printed in the United States of America.

In memory of my grandmother, Jean Lappley Young

Contents

	List of Illustrations	ix
	Acknowledgments	xi
	Introduction	1

Part 1: Ghana

1	The Disenchantment of Ghana's Basel Mission, 1828–1918	21
2	Enchanted Competition for the Presbyterian Church of Ghana, 1918–60s	53
3	The Enchantment of the Presbyterian Church of Ghana, 1960–2010	83

Part 2: North America

4	The School of Deliverance and the Enchantment of the Ghanaian Presbyterian Churches in North America	113
5	The Enchantment of the United Ghanaian Community Church, Philadelphia	136
6	Gendered Transformations of Enchanted Calvinism in the Ghanaian Presbyterian Diaspora	169
	Conclusion	193
	Appendix: Deliverance Questionnaire	209
	Bibliography	215
	Index	231

Illustrations

Figures

1.1	Odente shrine, 1899–1912	35
1.2	Child with amulets, 1881–1911	37
1.3	Pastor Kwabi, 1885–1911	45
1.4	Dr. Fisch's Aburi dispensary, 1897	50
2.1	Family processing cocoa pods on southern Ghana farm, 1920s	56
2.2	Anum Secondary School, 1907	72
3.1	Challenge Enterprises bookshop signboard, February 2007	93
3.2	Saturday deliverance service, Grace Presbyterian Church, Akropong, February 2007	99
3.3	Individual deliverance consultation, Grace Church, Akropong, February 2007	104
3.4	Individual deliverance consultation, Grace Church, Akropong, February 2007	105
4.1	New York deliverance workshop, November 2007	127
4.2	Drawing of biblical red dragon, New York deliverance workshop, November 2006	131
4.3	Woman possessed, New York deliverance workshop, November 2006	133
5.1	United Ghanaian Community Church congregation, January 2007	145
5.2	Flyer, Abboah-Offei's Philadelphia campaign, September 2005	150
5.3	Celebration at the United Ghanaian Community Church, January 2007	168

6.1 Rev. Rosamund Atta-Fynn praying for male congregant, December 2007 — 185

6.2 David Frimpong, installation as a deacon in the United Ghanaian Community Church, January 2007 — 188

Maps

I.1 Postcolonial Ghana — 3

2.1 Faith Tabernacle branches in greater colonial Ghana, 1918–26 — 63

4.1 Network of Ghanaian Presbyterian churches in North America, 2009 — 118

Acknowledgments

This book is based on ethnographic and historical research conducted in Philadelphia and New York as well as Accra and Akropong, Ghana, between July 2005 and July 2009. A number of people in each locale helped me to achieve my research goals. Special thanks to Dr. Rev. Kobina Ofosu-Donkoh, pastor of the United Ghanaian Community Church (UGCC) in Philadelphia, who generously met with me countless times for discussions and allowed me unlimited access to his church. Members of the UGCC's prayer team, particularly Isaac Baah, were also very helpful in allowing me insights into the thoughts and practices of their organization. In New York, Rev. Samuel Atiemo of the Ghanaian Presbyterian Church in Brooklyn helped me in numerous ways. In Akropong, Catechist Ebenezer Abboah-Offei and his wife, Faustina Abboah-Offei, kindly gave me a room and fed me during my stay in Akropong in February 2007. Catechist Abboah-Offei and Samuel Asare were very generous in allowing me to participate in the New York deliverance workshop in 2006, 2007, and 2008, as well as in the daily activities of Grace Presbyterian Center in Akropong. Thanks to Rosamund Atta-Fynn and Margaret Asbea-Oboagye for several formal and informal discussions about gender and women's roles in the Presbyterian Church in Ghana and the United States. Thanks to the congregation of the UGCC, Grace Presbyterian Church in Akropong, and all participants in the New York deliverance workshop for always making me feel welcome and including me in all activities.

Many people generously allowed me access to archival data that comprises chapter 2 of this monograph. Pastors Kenneth Yeager and Ronald Kilbride of Faith Tabernacle Congregation in Philadelphia provided me access to their church's archive. I am very thankful to Pastor Kilbride for his interest and enthusiasm in the history of Faith Tabernacle, which made for wonderful conversations both before and after services. Pastor Yeager kindly put me in touch with Faith Tabernacle of Ghana's Presiding Elder Samuel Okai and Elder

Solomon Quaye, who provided me access to their Ghana archive and arranged for me to conduct oral interviews with the oldest members of Faith Tabernacle in Ghana. Thanks to General Secretary Alfred Kuduah of the Church of Pentecost in Accra for providing me access to his church's archive. The staff of the Akrofi-Christaller Center in Akropong was very charitable in allowing me access to the archive of the Presbyterian Church of Ghana. Pastor Samuel Addai-Kusi of the Christ Apostolic Church in Accra also took the time to answer several questions about the early history of his church and gave me access to important historical documents. Finally, I owe a debt of gratitude to the helpful staff at the Philadelphia Presbytery, who granted me access to the UGCC file.

Many scholars and teachers helped me conceptualize my approach to this book. Foremost was Sandy Barnes, whose patience, loyalty, and commitment shaped my training, research, and writing. Steve Feierman, another mentor, significantly influenced my understanding of health and healing in religious contexts. Both Sandy and Steve demonstrated to me the importance of incorporating original historical research into a larger ethnographic project, which I attempted to follow in this book.

Benjamin Lawrance and John Peel also helped with various historical aspects, particularly in chapter 2. The support of Ann Matter, Michael Frechetti, Anthea Butler, and Joel Robbins has sustained this project from its inception. And in this vein, Stephen Prothero's enthusiasm for this book gave its final stages a big boost. Durba Chattaraj gave me insightful feedback on chapter 6, for which I am most appreciative. Sandy Barnes and Tom Callaghy helped me considerably with my conclusion, particularly in articulating my argument about the state in relationship to enchantment. Zeljko Rezek and Christine Murray, GIS experts both, were incredibly patient and thorough in constructing the three maps featured in this monograph. Josh Stanfield—an amazing copy editor—helped me to polish the final version of this monograph. Last but not least, Val Ross taught me how to write critically, and to her I am enormously grateful. Without Val's help and support this book would never have come to fruition.

The theoretical framework of this book, which significantly engages Max Weber's concept of religious enchantment in relationship to capitalism, developed over the course of two years, from 2009 through 2011. This engagement was in no small part a result of a series of extended conversations with Jeff Green during this period. Along with these conversations, Jeff's very insightful comments on the first draft of my introduction and the importance of his own article on Weber and the multiple meanings of enchantment gave this book the theoretical construct it needed. To Jeff, I owe a great debt of gratitude.

I gratefully acknowledge the financial support of the Department of Education Foreign Language Area Studies fellowships, via Penn's African Studies Center, that funded my language study and research for over four years. The Department of Anthropology at Penn granted me two stipends for summer field research in Ghana, and I received another grant from Penn's African Health Group for summer research, for which I am very thankful.

Thanks also to Koninklijke Brill NV for permission to reprint "Missionary Medicine and Akan Therapeutics: Illness, Health and Healing in Southern Ghana's Basel Mission, 1828–1918," *Journal of Religion in Africa* 39, no. 4 (2009): 429–61, which comprises the first chapter of this monograph. In addition, material within several chapters, but particularly chapter 6, first appeared in a chapter titled "School of Deliverance: Healing, Exorcism, and Male Spirit Possession in the Ghanaian Presbyterian Diaspora," which was published in *Medicine, Mobility, and Power in Global Africa: Transnational Health and Healing*, edited by Hansjorg Dilger, Abdoulaye Kane, and Stacey Langwick (Bloomington: Indiana University Press, 2013).

Thanks especially to the University of Rochester Press. Suzanne Guiod was a pleasure to work with as editorial director, and her enthusiasm was matched by that of Sonia Kane, who took over the position in the autumn of 2012. Toyin Falola, the series editor, was highly supportive from the moment I sent him the prospectus of this project. His support has been incredibly humbling and flattering.

To my family: thanks to my sister Jennifer Hyde, my mother Elizabeth Mohr, and my father Mark Mohr, for supporting my research. Special thanks to all my family in Sweden, who always provided laughter and relief in times of great stress. Thanks also to Susan and Tom Hyde—my sister's in-laws—who kindly hosted my family in their Williamstown, Massachusetts, home in December 2012, when I put the finishing touches on this book under the inspiration of the snow-covered Berkshires. Finally, my wife, Michaela, and son, Axel, had to endure the daily frustrations, pain, and joy that were involved in completing this monograph. To them I owe everything.

Introduction

In November 2006, at a rural Presbyterian retreat center located between New York City and West Point, a group of about one hundred Ghanaian immigrants from various Ghanaian Presbyterian churches in North American met for a week-long retreat. The event was announced as an occasion to "build capacity for pastoral care and service." This general description was used by the Ghanaians to explain the purpose of the gathering to their white, American, and religiously liberal Presbyterian hosts. These Ghanaians in fact were learning various techniques for the care of individuals within their congregations. But, more specifically, this group of Ghanaians from major North American cities was at this workshop to be formally trained by Catechist Ebenezer Abboah-Offei—the leading practitioner in the Presbyterian Church of Ghana's healing movement—in the theory and practice of Christian therapy. Among this network of Ghanaian Presbyterians, pastoral care was synonymous with deliverance, defined by Abboah-Offei as the process of freeing people as well as places and objects from the power of Satan and his demons.[1]

Since the mid-1990s a network of Ghanaian Presbyterian churches has been growing in North America. In the midst of the stress and uncertainty involved in the process of labor migration,[2] spiritually defined disorders were afflicting this community. Partnering with Catechist Abboah-Offei, North American church leaders decided to

---

1. See Asamoah-Gyadu, *African Charismatics*, 164–200; Gifford, *Ghana's New Christianity*, 83–112; and Meyer, *Translating the Devil*, 141–74 for discussions of deliverance practices within Ghanaian Christianity.

2. By labor migration, I am primarily referring to the migration of working-age people who intend to work higher paying jobs in the United States and Canada. The majority of Ghanaians in the United States are of working age: 74 percent of the Ghanaian-born immigrants in 2000 were between the ages of twenty-five and fifty-five (US Census Bureau). Many immigrants also migrate for educational purposes, with the intent as well of obtaining higher-paying jobs. I include these educational migrants in the larger category of labor migrants. The great majority of Ghanaian migrants since the mid-1990s are neither refugees nor asylees.

train groups called "prayer teams" within each Ghanaian Presbyterian congregation in North America. Prayer teams are subchurch organizations responsible for spiritually managing the health and welfare of the churches' congregations.

Because Abboah-Offei could not attend in 2006, his assistant from Ghana, Samuel Asare, led the workshop. During several lectures given that week, Asare frequently related the suffering of the Israelites in their flight from Egypt into the Promised Land to the travails of the Ghanaian Presbyterians in their journey from West Africa to North America. In reference to both migrations, Asare consistently highlighted the point that "their journeys were not without demonic confrontation." During the course of this workshop several spiritual attacks occurred: in one, a man was afflicted by seizures thought by many to be the work of witches; in another, a fire in a pregnant woman's room was interpreted to be the result of a biblical dragon working in conjunction with witches to consume her unborn baby.[3] Besides prayer-team members, several afflicted Ghanaians from greater New York and beyond flocked to see Samuel Asare in order to be treated for a host of illnesses and misfortunes that afflicted Ghanaians in North America.

At the turn of the twenty-first century, these Presbyterian labor migrants from Ghana led enchanted lives. In this book, I will consider how and why Ghanaian Presbyterian communities, which were formed in the first half of the nineteenth century, became *more* enchanted the more they integrated into capitalistic modes of production, particularly in the context of labor migration. This relationship of capitalism to enchanted Calvinism (of which Presbyterianism is one type) is the inverse of that posed originally by Max Weber, who argued that capitalism was disenchanting Calvinism at the turn of the twentieth century.

Capitalism, Calvinism, and Disenchantment

Max Weber argues that the spirit of modern capitalism, whereby individuals continually reinvest profits into their businesses, results from a religious orientation found in early Calvinism. Weber makes this case most prominently in *The Protestant Ethic and the Spirit of Capitalism*,

---

3. For a more detailed discussion of these events, see chapter 4.

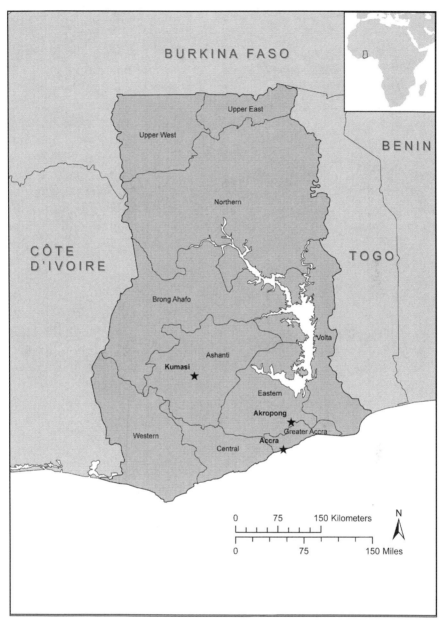

Map I.1. Postcolonial Ghana. Map by Christine Murray.

published originally as a two-part article in 1904 and 1905. He begins his discussion with the observation that business leaders, owners of capital, and the most skilled labor—specifically in his home province of Baden but by extension Western Europe and North America—are overwhelmingly Protestant. Furthermore, many of the upper ranks of the business world are also indifferent to religion. Weber's argument, designed to explain this observation, is summarized as follows.

A certain version of worldly asceticism developed within Protestantism and became a significant social force in Western Europe and the United States. This idea began with Martin Luther's conception of "the calling," which refers to the idea that the highest form of moral obligation for a Protestant is to fulfill his or her duty in worldly affairs. This idea gave religious significance to action in this world, where labor became the outward expression of brotherly love.[4]

This conception of the calling became more rigorously developed in various forms of Protestantism, which Weber refers to as the Puritan sects,[5] such as Methodism, Pietism (a movement within the German Lutheran Church), and the Baptist movement, but most significantly within Calvinism. Particularly important to Weber's argument about Calvinism was the idea (developed by John Calvin) of predestination: certain people were predetermined by God to be saved from damnation. Salvation did not depend on what a person said or did. Instead, one's chosen status was indexed through morally righteous action in this world that became the sign—not the means—of one's election within Calvinist communities.

How did this religious idea affect modern forms of capitalism? The accumulation of wealth was morally sanctioned within Calvinism only if combined with an industrious career devoid of self-indulgence. Worldly pleasures were not to be enjoyed. Unlike Catholicism, a rejection of worldly pleasure through monasticism was not the ideal. Calvinists were to work hard and be successful in order to signify their chosenness, but they could not enjoy the fruits of their labor. Instead, their profits were to be accumulated

---

4. Weber, *Protestant Ethic*, 81.

5. Weber is not using the term "sect" in the more contemporary pejorative sense. By sects, Weber means Protestant denominations that have separated from state-sponsored churches. For instance, Methodism separated from the Church of England and Pietists in Germany separated themselves in many ways, although not establishing a separate denomination, from the Lutheran church.

and reinvested in their business. This, argues Weber, is the ethos of modern capitalism, which originated within Calvinist communities in Western Europe and the United States.

The growth and spread of modern capitalism, however, contributed to the disenchantment of the world. The disenchantment of the world—as a condition of modernity[6]—is a major theme within Weber's work, although rarely discussed or thematized directly. In fact, Weber never explicitly refers to disenchantment in the *Protestant Ethic*. What exactly is meant by disenchantment? As Jeffrey Green demonstrates, there are several senses in which Weber uses disenchantment throughout his writings.[7] For my purposes, I will focus on the religious sense of disenchantment.

Derived from the German term *Entzauberung*, which most literally translates as "de-magicization," religious disenchantment refers to three things.[8] One, disenchantment refers to the decline of supernatural modes of explanation from the world and their replacement with worldly explanations. Two, disenchantment refers to the departure of supernatural forces behind the natural world that interact in human affairs, such as spirits, demons, and gods. Three, disenchantment refers to the absence of charismatically endowed humans—referred to alternatively as sorcerers, magicians, and spiritual advisors—to manipulate these supernatural beings in accordance with human agency.

Weber argues that two interrelated social forces produced this form of disenchantment: capitalism and science.[9] Capitalism's religious foundation was destroyed as capitalism expanded, which corresponded to the rationalization of religion within the Protestant sects. Financial accumulation via a capitalist economy became an end in itself, losing its prior religious and ethical meaning. And no one could escape capitalism's reach; those who could not adapt to

---

6. Modernity is ultimately an ideology of history: a belief or attitude about social practices, viewed in terms of a radical historical discontinuity with the past (Agha, "Speciation of Modernity").

7. Green, "Two Meanings of Disenchantment."

8. Weber, *From Max Weber*, 139.

9. While I treat these two phenomena separately, they were interrelated. As Weber argues, the development of science enabled modern capitalism: capitalism is a product of economic rationality, defined as the expansion of productivity by the scientific restructuring of production (*Protestant Ethic*, 24).

capitalism were doomed to fail.[10] Wealth, in essence, had a significant secularizing effect, leading to the general decline of spirituality.[11]

In Weber's view, the development of modern science was another significant force in the religious disenchantment of the modern world. Scientific methods led to a much greater control over the natural world than was possible with magic or religion, because natural phenomena became reducible to calculation.[12] With calculable control, science replaced much religious explanation of the natural world, causing religion itself to be seen as magical, superstitious, and above all, irrational.[13] In particular, the medical advances during the nineteenth century had a considerable disenchanting effect on Euro-American society. Within clinical medicine, William Morton (1846) developed anesthesia techniques that made surgery painless, and Joseph Lister (1865) began disinfecting surgical equipment, thereby making operations safer. Furthermore, during Weber's lifetime (1864–1920), the bacteriological revolution unfolded in the 1880s. This medical revolution was constituted by a series of discoveries of specific causal agents of infectious disease that transformed medical practice and theory, most notably by Louis Pasteur and Robert Koch. This includes the discovery of the bacteria causing typhoid (1880), pneumococcus (1881), tuberculosis (1882), cholera (1883), and diphtheria (1884). Once the bacteriological agent was discovered, the corresponding diseases were amenable to treatment with antibiotics.[14] Disease was shown to result from nature, not evil.[15]

Ideal Types, Heterodoxy, and Enchanted Puritanism

In his writings about Calvinism and disenchantment, Weber relies on an orthodox view of Protestantism, derived from a Kantian framework that juxtaposes religion as cult to religion as moral action. For Immanuel Kant, a cult seeks favor from God through prayer and

---

10. Weber, *Protestant Ethic*, 54–55.
11. Ibid., 174–83.
12. Ibid., 24.
13. Weber, *From Max Weber*, 351.
14. While scientific advances did have a secularizing effect on society, scientific advances, at times, did lead to the enchantment of health science, hence the popular term "magic bullets" for antibiotics. Michael Saler argues that modern science has become a central locus of modern enchantment (*Modernity and Enchantment*, 714).
15. Foucault, *Birth of the Clinic*, 196.

offerings to bring health, among other things, to its followers (i.e., enchantment), while religion as moral action commands human beings to change their behaviors in order to lead better lives (i.e., the Protestant ethic).[16] Furthermore, Weber believes that a demand from religion for security, health, and well-being in this world, rather than a meaningful life or spiritual perfection, was a corruption of the religious drive.[17]

Orthodox forms of Protestantism described by Weber did not have a monopoly over Protestant practice during the nineteenth and early twentieth centuries, however. Heterodox forms of Protestantism—groups that did believe in modern-day miracles and, in particular, healing miracles—flourished within certain segments of Euro-American society. These heterodox forms of Protestantism and healing do not have a place in Weber's writing, which focuses on the most orthodox forms that he calls the "ideal type."

Weber uses ideal types to describe the most orthodox form of any given category.[18] The ideal type is an abstraction, used as a standard of comparison, which enables a reader to see aspects of the real world in a clearer, more systematic way. For example, by comparing the ideal type of democracy with actual democratic societies, a researcher can judge how these societies correspond or depart from the ideal type.

Ideal types have drawbacks, however. One problem is that the act of defining and delineating ideal types—that which constitutes orthodox Protestantism, for instance—is an act of social creation. As Pierre Bourdieu argues, there is power embedded in this taxonomic process enacted by professional theoreticians to name or define groups, thereby affecting or constituting the public perception of group orthodoxy.[19]

---

16. Kant, *Religion within the Boundaries of Mere Reason*.
17. Turner, *Religion and Modern Society*, 287.
18. Weber, *Protestant Ethic*, 98.
19. Bourdieu, "Social Space." The far-reaching consequences of Weber's defining orthodox Protestantism as ascetic have affected the way in which anthropology as a discipline has interpreted Christianity: above all else, as containing an antagonism between body and spirit, therefore dealing poorly with issues like illness that pertain to the body (Cannell, "Introduction," 39). When Christianity, particularly in the global South, has been observed by anthropologists to include bodily oriented practices, as with healing and deliverance, the faith has been frequently judged to be syncretic at best, illegitimate at worst. It is this trend in anthropology that I wish to

Another drawback is that ideal types produce a rigidity that tends to suppress the range of variation within a given type, at times homogenizing any given phenomenon that it constitutes. This typological rigidity can furthermore conceal the existence of heterodox components, which may account for gradual, cumulative sociocultural change.[20] The heterodox ideologies and practices that are disregarded by Weber, I argue, are essential to understanding the state of Calvinism, capitalism, and enchantment today; that which was heterodox at the turn of the twentieth century has become significantly more orthodox at the turn of the twenty-first century. This process has occurred gradually over time.

While Weber's usage of ideal types is problematic in creating typological rigidity, he does concede (in a footnote) that ideal types do violence to historical reality in a certain sense.[21] He qualifies his ideal types, at times, with data (particularly located in extensive footnotes) situated outside the boundaries of a given ideal type, thereby creating a certain tension in his work between orthodox representation and heterodox practice.

One example of this tension is found within one of Weber's central propositions: that Protestantism at the turn of the twentieth century was disenchanted. Contrary to this argument, Weber claims that not even among Protestants had the existence of spirits and demons permanently disappeared.[22] Their suffering was managed via pastoral care, which was the rationalized form of prophetic religion.[23] Pastoral care may assume diverse forms, claims Weber, and can be closely assimilated to "magical manipulations": the use of supernatural agency to resolve elements of tension in life situations.[24] Here we see all three aspects of enchanted Protestantism: spiritual explanations of the world, spirits interacting with people in this world, and

---

transcend in this study by examining the ways in which Ghanaian Presbyterians have engaged various heterodox forms and practices of Christianity over time and space.

20. Parsons, "Introduction," lxxvi.
21. Weber, *Protestant Ethic*, 233.
22. Weber, *Sociology of Religion*, 20.
23. Ibid., 75–76. The normative way to deal with suffering within twentieth-century Protestantism—interpreted as a blessing permitted or even ordained by God—was patient endurance (Curtis, *Faith in the Great Physician*, 15). This practice was influenced by the Calvinist interpretation that God had ceased to work miracles—divinely interacting with humans on their behalf—after the apostolic era.
24. Parsons, "Introduction," lvii; Weber, *Sociology of Religion*, 75.

charismatically endowed humans managing these spirit-human interactions. Weber, however, cites no specific examples.

The most enchanted form of Protestant pastoral care in the nineteenth century was linked to Ghana through mission. This form of enchanted Protestantism that combated the powers of Satan and his demons, was found in German Pietism—the so-called fourth generation—which emerged during Weber's lifetime. In Württemberg—adjacent to Weber's home province of Baden—the Pietism of Johann Christoph Blumhardt, who focused on healing the body and soul as well as casting out demons from the spiritually possessed, flourished in the mid-nineteenth century. Württemberg Pietism's missionary wing, the Basel Mission, was the missionary predecessor to the Presbyterian Church of Ghana (see chapter 1).

While this Protestant healing movement began to wane in Germany after Blumhardt's death in 1880, it did not disappear, but became popularized primarily in England and the United States. Called divine healing or faith cure, this form of Protestant healing was first practiced within the Methodist Holiness and Calvinist Higher Life movements in the last quarter of the nineteenth century,[25] followed by Pentecostalism, which emerged in the first decade of the twentieth century.[26] Ghanaians affiliated with the Presbyterian Church—and interacting significantly with missionaries and publications from the divine healing and Pentecostal movements—participated in and interpreted these Christian healing movements in a variety of ways that shifted over space and time (see chapter 2). By the 1960s and into the twenty-first century, Christian healing practices became popularized even within Calvinism, including in the Presbyterian Church of Ghana, thereby constituting its enchantment (see chapters 3–6).

---

25. Divine healing Calvinists such as the Presbyterian William Boardman, Presbyterian Albert Benjamin Simpson, and Congregationalist John Alexander Dowie were all leaders in the divine healing movement.

26. Several particular features of Blumhardt's ministry were adopted within the faith cure and Pentecostal movements. Tensions and debates between the medical and Christian communities over the legitimate right to heal continued, particularly with respect to Christian healing homes (Curtis, *Faith in the Great Physician*, 136–39). The healing home created by Blumhardt at Bad Boll was frequently replicated by divine healing and Pentecostal advocates. Blumhardt was frequently remembered as the modern progenitor of divine healing in print through the production of biographical literature and within orations at healing and holiness conferences.

Enchanted Calvinism, therefore, refers to the institutionalized recognition of spiritual explanations of the world by the Presbyterian Church of Ghana, the institutional acknowledgment of spiritual afflictions in the lives of the church's congregants, and the decisive ways in which these spiritual afflicted individuals were and continue to be delivered from their maladies by charismatic individuals supported by the church. This book is about spiritual affliction and spiritual healing within the Presbyterian Church of Ghana, particularly under the conditions of labor migration: first, in the early twentieth century during the cocoa boom in Ghana, and second, at the turn of the twenty-first century in the context of the healthcare migration from Ghana to the United States. I argue that the more these Calvinists became dependent on capitalist modes of production, the more enchanted their lives, and subsequently their church, became, although in different ways and taking on different forms within these two migrations. The enchantment of a particular branch of Calvinism corresponds to a significant global shift of Christianity generally and Calvinism specifically.

Global Shifts of Christianity and Calvinism

The center of Christianity, based in Europe and North America in the twentieth century, has begun to shift rapidly to the global South. In 1900, 80 percent of the world's Christians lived in Europe and North America. By 2000, 60 percent of the world's Christians lived outside of Europe and North America. This shift in Christianity is predicted to be even more pronounced throughout the twenty-first century, which will be one of the most transformative centuries in the history of world religions.[27]

Africa, and African Christianity, is at the center of this transformation: the large-scale adoption of Christianity being one of the master themes in African history during the twentieth century.[28] In 1910, there were just over 8,500,000 Christians in Africa. One hundred years later, in 2010, the estimated number of African Christians was over 516 million.[29] In 1910, 9 percent of sub-Saharan Africa's

---

27. Jenkins, *Next Christendom*; Walls, "Eusebius Tries Again."
28. Peel, *Religious Encounters*, 1.
29. Pew, *Global Christianity*, 15.

population was Christian; in 2010 that number rose to 63 percent.[30] Since the 1980s African immigrants in North America and Europe have been establishing African immigrant churches that have in turn been transforming the Christian landscape in Euro-America.[31]

Within African Christianity, Pentecostal-type Christianity has come to define the faith. Even more specifically, healing has become a distinguishing feature and perhaps *the* distinguishing feature of African Christianity compared with other regional Christianities.[32] Therefore, a specific type of spirit-centric, healing-focused Christianity is growing both within Africa and among African migrants in the global North. Far from its disenchanted missionary roots at the turn of the twentieth century, African Christianity has become enchanted in the twenty-first century.

This transition includes the Calvinist or Reformed churches, of which Presbyterians comprise the largest subset. Today, the global landscape of the largest Calvinist Churches is also outside of the Euro-American heartland of 1900. Of the twenty-one Calvinist Churches with over five hundred thousand members that are affiliated with the World Council of Churches—the largest ecumenical organization in the world—only five reside in Europe or North America.[33] Only three of these five have over one million members: the Presbyterian Church USA (2,501,181), the Federation of Swiss Protestant Churches (2,416,973), and the Church of Scotland (1,149,000).

The great majority of the largest Calvinist denominations, sixteen out of twenty-one, are located in Asia and Africa. In Asia, the Presbyterian Church of Korea claims 2,852,311 members, while Indonesia is home to four significantly large Calvinist churches: the largest being

---

30. Ibid., 10.

31. Ludwig and Asamoah-Gyadu, *African Christian Presence*.

32. Cox, *Fire from Heaven*, 254–57; Jenkins, *Next Christendom*, 123–27. Jenkins argues that the ways in which spiritual forces affect the lives of believers is the single area of belief and practice that divides Christian orthodoxy in the global South from the global North.

33. The following statistics on Calvinist/Reformed churches is found on the World Council of Churches (WCC) website: www.oikoumene.org/en/member-churches (accessed April 23, 2013). These statistics were provided to the WCC by each member church, with no common criteria for establishing membership numbers. The figure for the PCUSA on the WCC website, over 2,500,000, is significantly more than the most recent statistics given on the Presbyterian Church of the USA website for 2009, 2,077,000. The PCUSA's membership has not been above 2,500,000 since 2000.

the Protestant Evangelical Church in Timor (two million).[34] Africa, which comprises more than half of all Calvinist denominations of over five hundred thousand members, is at the center of global Calvinism today. Seven of these churches have over one million members, the largest being the Presbyterian Church of East Africa based in Kenya (four million).[35] Among the four African Calvinist denominations with between five hundred thousand and one million members is the Presbyterian Church of Ghana (565,637), which is the focus of this monograph, along with its branches among the Ghanaian diaspora in North America.[36]

The spread and extraordinary growth of Calvinism in Ghana, its enchanted quality, and its development in North America were certainly unintended and unimaginable consequences of the late nineteenth- and early twentieth-century Calvinism examined by Weber.[37] Weber's historical explanation in *The Protestant Ethic* and other writings is distinctly about the religious-ethical origins of modern capitalism in Western Europe and North America. He does not make suggestions about the spread of Christianity, capitalism, or disenchantment elsewhere.

In this book, alternatively, I examine the relationship between Calvinism, capitalism, and enchantment in Ghana and the Ghanaian diaspora in North America. More specifically, I consider how and why

---

34. The others being the Christian Evangelical Church in Minahasa (730,000), the Evangelical Christian Church in Tenah Papua (600,000), the Protestant Church in Western Indonesia (600,000), and the Protestant Church in the Moluccas (575,000).

35. The other African Calvinist denominations with more than one million members are the Presbyterian Church of Nigeria (3,806,690); the Presbyterian Church of Africa, which is based in South Africa but includes members from Malawi, Zambia, and Zimbabwe (3,381,000); Church of Christ in Congo—Presbyterian Community of Congo (2,500,000); Evangelical Church of Cameroon (2,000,000); Presbyterian Church of Cameroon (1,800,000); and the Presbyterian Church of Sudan (1,000,000).

36. The other three largest African Calvinist churches with between five hundred thousand and one million members are the Evangelical Congregational Church in Angola (950,000); the Church of Central Africa, Presbyterian [Malawi] (770,000); and the Presbyterian Church in Cameroon (700,000).

37. The study of unintended consequences was a driving force behind much of Weber's research (see Weber, *Max Weber*, 337). In particular, Weber focuses on the ways in which ideas and material interests can influence one another, with the causal chain switching directions through time owing to a variety of forces on social life (280). Weber sees the causal chains running in divergent and occasionally inverse directions through time.

Ghanaian Presbyterian communities, which began to be organized in the first half of the nineteenth century, became significantly *more* enchanted the greater they became integrated into capitalistic modes of production, particularly in the context of labor migration.

Outline of Chapters

This book is divided into two parts, both containing historical and ethnographic components. The first part focuses on the Presbyterian Church of Ghana from 1828, when the first Basel missionaries arrived in Ghana, to the (ethnographic) present in Ghana, where I conducted research for a total of five months over three trips, in the summer of 2001, the summer of 2003, and in February 2007. The ethnographic data were collected in Accra and Akropong. The historical data were collected from the Basel Mission Archive in Basel, Switzerland, the Faith Tabernacle Archive in Philadelphia and Accra, the Church of Pentecost Archive in Accra, and the Presbyterian Church of Ghana Archive in Akropong.

The second part of the book focuses on the Ghanaian Presbyterian immigrant community in North America: the United States primarily, but I frequently include the Canadian community in my discussion. I refer to "Ghanaian Presbyterians in North America" when the Canadian congregations are included in the discussion, particularly in chapter 4. Otherwise, I refer to "Ghanaian Presbyterians in the United States." Ethnographic research for the second part was conducted in Philadelphia from the summer of 2005 through the summer of 2009 as well as during three week-long trips to New York in November of 2006, 2007, and 2008 to participate in the deliverance workshop that is the subject of chapter 4. Historic data were collected at the archive of the Philadelphia Presbytery as well as orally from leaders of the Ghanaian Presbyterian community in North America.

In chapter 1, I trace the establishment of Ghana's Basel Mission, the institutional predecessor to the Presbyterian Church of Ghana. From the mission's beginnings until around 1885, both European missionaries and Ghanaian Christians followed a local pattern of therapeutic pluralism, consulting several types of traditional healing practitioners for relief from sickness and misfortune. While not an explicit part of mission theology or practice, the Basel Mission was informally enchanted in Ghana before 1885. The missionaries and Ghanaian Christians shared spiritual interpretations of the world and believed that malevolent spiritual forces afflicted Christians. The Mission, however, provided no form of spiritual healing, but instead outsourced its healing practices to various Akan healers.

As biomedicine in Europe became more of a rationalist science at the turn of the twentieth century, Basel missionaries began to shun traditional therapeutics in favor of biomedicine. This process began around 1885, when the first Basel medical missionary arrived in Ghana. After this time, the mission denied the existence of malevolent spiritual forces and failed to continue supporting the use of Akan healers. Ghanaian Christians, however, frequently subverted the authority of European missionaries and continued consulting traditional healers, while the Basel missionaries became more convinced that these therapeutic practices were demonic. This subversion on the part of Ghanaian Basel Christians within the institutional framework of the church continued steadily until 1918.

The second chapter begins by recounting the rise of the cocoa industry in Ghana, which was the largest producer of cocoa in the world by 1918. This lucrative industry—managed by local Ghanaians and not expatriate European firms—first emerged within the Basel community and was dominated, in its earliest years, by Ghanaian Basel Christians. Many young Basel Christians migrated great distances to farm cocoa, an industry that wreaked social havoc on Ghanaian communities as capitalism disrupted traditional forms of social organization. This disruption was frequently expressed through the idiom of witchcraft, which particularly threatened migrant cocoa farmers.

The year 1918 marked the beginning of great change for the Basel Mission[38] and its relationship to Christian therapeutics. Three interrelated events occurred that year: the (mostly German) Basel missionaries were expelled from British-controlled Ghana as a result of the conflict in Europe, the 1918–19 influenza pandemic erupted, and the cocoa market crashed. These three events caused a crisis within the Ghanaian Basel community, particularly among the migrant cocoa farmers. As a result of this crisis, many left the church

---

38. After 1918, when the Basel missionaries were expelled, the mission was colloquially referred to as the Scottish Mission. By 1926, the Scottish Mission formally became the Presbyterian Church of the Gold Coast and further changed its name to the Presbyterian Church of Ghana after independence in 1957.

and joined an emergent divine healing church called Faith Tabernacle Congregation, which offered robust Christian healing practices to combat both the influenza epidemic and socio-spiritual disorders attributed to witchcraft. Between 1930 and 1960 young Presbyterians continued to leave their churches in favor of newer Pentecostal churches—many of them branches of Faith Tabernacle—that offered Christian therapy.

Chapter 3 begins by examining the Presbyterian Church of Ghana's response to the mass exodus of its members after independence in 1957. By the early 1960s, the Presbyterian Church's leadership decided to formally institute healing practices in the church. This decision culminated in the mid-1990s with the establishment by Catechist Ebenezer Abboah-Offei of Grace Presbyterian Church, a charismatic church also referred to as a deliverance center, which formally combated spiritual disorders that plagued members of the Presbyterian Church. By this time, the Presbyterian Church of Ghana had become enchanted. The context of the Presbyterian Church of Ghana's enchantment was significant political, economic, and medical decline in Ghana.

Chapter 4 describes the establishment of a network of Ghanaian Presbyterian churches in North America among Ghanaian migrants who predominately found employment within the healthcare industry. This generation of Ghanaian Presbyterian labor migrants, like their grandparents who migrated to the cocoa farms, experienced spiritual afflictions frequently attributed to witchcraft. Unlike their grandparents, however, their needs were met within the Presbyterian Church of Ghana, when Catechist Abboah-Offei began evangelizing in North America and establishing healing practices among the diaspora churches. This chapter focuses on the primary site of replicating Christian healing practices (developed at Grace Presbyterian Church), which, since 2004, has been the New York deliverance workshop that trains deliverance practitioners from each of the Ghanaian Presbyterian churches in North America.[39]

The growth and establishment of deliverance practices in one particular Ghanaian Presbyterian Church in the United States—the United Ghanaian Community Church (UGCC) in Philadelphia—is the focus of chapter 5. Similar forces influenced the creation of

---

39. The deliverance workshop in 2012 was held in southern New Jersey alternatively.

healing practices within the UGCC as in the Presbyterian Church of Ghana—spiritually suffering members as well as competition for those members from Pentecostal churches, particularly the Church of Pentecost. Ten years after the establishment of the UGCC in 1995, a group of healing practitioners was organized—with significant help and stimulus from Abboah-Offei—to counsel, diagnose, and deliver the afflicted within their congregations. By 2006 this entire cohort of healing practitioners was voted into leadership positions within the UGCC—as elders and deacons—ensuring a central place of Christian therapy within the political structure of the congregation.

Chapter 6 examines disjunctures in healing and deliverance practices that have occurred in the context of migration to North America. In particular, gender relationships have been significantly altered within the Ghanaian Presbyterian immigrant community. Because of the greater earning capacity of Ghanaian women in the traditionally feminized US healthcare industry, many Ghanaian women earn more money than their husbands. The greater economic independence of Ghanaian women in the United Staes correlates with a shift in female empowerment in the Ghanaian Presbyterian churches and in the deliverance ministry particularly. In the United States, Ghanaian women are taking leadership roles within the church and are working as deliverance counselors, which is very uncommon in Ghana. Conversely, there is a greater need for healing and deliverance among Ghanaian men as demonstrated by the high rate of male spirit possession, something rarely seen in Ghana. This high rate of male spirit possession is an index of an emergent gender category, which I have termed "junior male femininity."

In the conclusion, I address two lingering questions in light of the developed proposition of this book. First, I explain why biomedicine has not been a disenchanting force in Ghana. There are at least five reasons for this, all of which relate to problematic understandings of and experiences with biomedicine in Ghana. One, biomedicine became associated with sorcery and, correspondingly, physicians became associated with sorcerers. Two, biomedicine was, at times, understood to be violent. Three, biomedicine, under certain conditions, was perceived to be exclusionary. Four, biomedical forms of economic transactions were believed to violate a local moral economy. And a fifth reason, which is more materially focused as opposed to cultural-critical, is that access to biomedicine was and continues to be very limited in Ghana.

Finally, I examine how politics, or the state specifically, affects the relationship between capitalism and enchantment. Here, I make two arguments about the state in relationship to religious enchantment. One, I claim that state welfare spending has an inverse relationship to religious enchantment. Low levels of state welfare spending in Ghana have therefore contributed to higher levels of religious enchantment. Two, I contend that the form religious enchantment takes in Ghana is affected by the rural land tenure system. Corporate landholding by extended families in Ghana maintains high levels of social expectations of reciprocity within families, which, when not met by many labor migrants, frequently results in socio-spiritual afflictions perpetrated by extended family members in Ghana.

Part 1

Ghana

1

The Disenchantment of Ghana's Basel Mission, 1828–1918

This chapter examines the healing practices that developed within Ghana's Basel Mission community between 1828, when the Basel Mission was first established in Ghana, to 1918 when the German and Swiss Basel missionaries were expelled from the British colony. The 1880s, however, was the most transformative decade of the first ninety years of Ghana's Basel Mission with respect to health and healing practices within the mission. Prior to the 1880s, the Basel Mission in Ghana was partially enchanted, while after the 1880s, the Basel Mission became institutionally disenchanted. This chapter explains the therapeutic transformation within Ghana's Basel Mission that occurred in the 1880s—and the crisis it produced.

The Development and Growth of the Basel Mission, 1828–1918

The Basel Evangelical Missionary Society (*Basel Missionsgesellschaft*) was the missionary arm of Württemberg Pietism.[1] The Mission emerged from the German Society for Christianity (*Deutsche Christentumsgesellschaft*) as a Bible study and discussion group created in 1780 that brought together prominent professionals within the Pietist movement. The members founded the Basel Mission in 1815 as a seminary for the education of overseas evangelists. The first graduates were not sent out as Basel missionaries, but joined older established evangelical missions such as the Dutch Missionary Society, the North German Mission Society, and the Church Missionary Society, with which

---

1. The Inspector or Director of Mission was always an ordained Württemberg Pietist, and at any given time almost half of the missionaries were from Württemberg (Smith, *Presbyterian Church*, 21).

Basel kept particularly close ties.[2] By 1821, the founders decided that the Mission must establish its own religious outposts abroad to bring the distinctive Pietist worldview to the un-Christianized world. In December 1828, at the invitation of the Danish crown and the Danish Lutheran Church, the first team of Basel missionaries arrived at the Danish trading settlement of Christiansborg in Ghana.

The early years of the Basel Mission in Ghana, from 1828 to 1850, were marked by a high mortality rate of European missionaries and a low conversion rate among Ghanaians. In the first four years of the Mission, from 1828 to 1832, seven missionaries were sent to Ghana and only one, Andreas Riis, survived. Not one Ghanaian convert was made during this time; therefore, Riis decided to start a Basel Mission station elsewhere. On March 19, 1835, Riis founded a Basel Mission station in Akropong, the capital of the Akuapem Kingdom, located twenty-five miles northeast of Accra. Akropong was chosen for three reasons: a lower incidence of malaria due to the high elevation, a population less influenced by amoral European traders, and to escape Ghanaians' suspicions that the Basel missionaries were political agents of the Danish, who at that time controlled Christiansborg.[3] Converts were not easily made in Akropong either. By 1840, twelve years had elapsed since the first Basel missionaries arrived in Ghana and still not a single Ghanaian convert was made. Meanwhile, eight of the nine Basel missionaries who arrived in Ghana had died.

In 1843 a new strategy was put forth by the Basel Mission to try to gain converts. The mission decided to invite West African-descended Jamaicans to Ghana in order to demonstrate that Christianity was also a religion for people of African decent. The Jamaicans landed at Christiansborg on April 17, 1843, and in the following June they moved to Akropong. More Basel missionaries arrived from Europe through the 1840s. By 1851, the Basel Mission had stations in Christiansborg, Akropong, and Aburi, and just twenty-one Ghanaian converts.[4] Early Ghanaian converts, particularly in Akropong, tended to fall into a few main categories of persons with ambiguous social status, such as princes (who would not become rulers), ex-slaves (who could escape servitude through conversion), divorced or deserted

---

2. Jenkins "Church Missionary Society"; Macchia, *Spirituality and Social Liberation*, 39–40.

3. Smith, *Presbyterian Church*, 30.

4. Miller, *Missionary Zeal*, 23.

women, orphans, and various kinds of people who had broken politico-religious taboos.[5] Christianity offered a way out of various existing constraints and promised upward social mobility for the dispossessed. Through the second half of the nineteenth century, the number of converts began to grow.

The period from 1850 to 1870 was marked by an accelerated growth rate of the Christian community due to a mastery of Twi by certain missionaries, evangelical activities in the major Akuapem towns around Akropong, training of Ghanaian personnel, and, more generally, the development of trade and agriculture.[6] The most significant growth in the Christian community came from the Basel Mission schools, as parents thought it advantageous to send their children to European-model schools. The Basel Mission schools taught students in Twi primarily, but also taught English, as the highly sought-after clerical positions in the colonial government required the use of spoken and written English.[7] By 1869, there were eight main mission centers with sixteen small Christian groups attached thereto, for a total Christian community of just over two thousand.[8]

The church continued to grow steadily into the early twentieth century. After 1875, the Basel Mission concentrated its efforts on church building in Ga, Krobo, and Akuapem areas and on establishing a church in Asante. In January 1896, the British military marched on Kumasi, the capital of Asante, and defeated its army. British control of Asante allowed Basel missionaries to establish a station in Kumasi in June 1896.[9] By 1898, the Basel Mission had 157 congregations, 21 ministers, and 128 schools with nearly 5,000 students; comprehensively, it was a Christian community of 16,806 members in Ghana.[10] At this time there were nearly 200 Ghanaian pastors, teachers, and other Basel Mission employees.[11] Within this emerging religious community, men were given positions of authority much more than women, ultimately contributing to a general increase in patriarchy in Ghana during the nineteenth century.

---

5. Middleton, "One Hundred and Fifty Years," 4.
6. Smith, *Presbyterian Church*, 45.
7. Zimmerman, *Alabama in Africa*, 125.
8. Ibid., 108.
9. By 1914 there were twenty Basel congregations, eight hundred converts, and seventeen Basel schools in Asante (Smith, *Presbyterian Church*, 130).
10. Ibid., 292.
11. Jenkins, "Afterward," 198.

But by the mid-1910s political conflict between British and Germans during World War I facilitated the expulsion of the German and Swiss Basel missionaries. By 1916 the British authorities in Ghana began to view the Basel missionaries, who were predominantly German, with suspicion. In the second week of December 1917, all of the German missionaries were brought to Accra and deported on the sixteenth. On February 2, 1918, the secretary of state in London ordered all remaining Basel missionaries home. The colonial government invited the United Free Church of Scotland to assume responsibility for the Basel missionary work.[12]

By the time the Basel missionaries were deported in early 1918, the church had grown considerably since its inception ninety years earlier. It claimed more than ten central stations and nearly two hundred smaller Basel congregations, which were led by thirty Ghanaian pastors along with several catechists and teachers. The total Basel community in 1918 was thirty thousand.[13] After 1918, the Ghanaian leadership made all major institutional decisions, and by 1926 a new name—the Presbyterian Church of the Gold Coast—was adopted (and later changed to Presbyterian Church of Ghana at the time of independence in 1957).

Basel, the missionary wing of Württemberg Pietism, was one of its four major centers in the nineteenth century, along with Stuttgart, Korntal, and Bad Boll.[14] Bad Boll became significant because it was the center of Württemberg healing practices, where Johann Christoph Blumhardt's Christian healing home or *Kurhaus* was located. In the following section I will describe the establishment and significant features of this nineteenth-century Pietist healing movement based in Bad Boll.

Christian Healing in Württemberg Pietism, 1800–1880

In the early nineteenth century, a strand of Pietism emerged in Württemberg, led by Christian Gottlieb Blumhardt, that focused on the liberation of the sick from the devil. The existence of sickness, as well as

---

12. This mission was led by Rev. A. W. Wilkie and Rev. J. Rankin, who were missionaries in Calabar, Nigeria. Wilkie took the step of organizing the Basel Mission as a self-governing Presbyterian Church.

13. Smith, *Presbyterian Church*, 154.

14. Macchia, *Spirituality and Social Liberation*, 22.

antisocial or immoral behavior, was explained by reference to Satan. These themes within Christian Blumhardt's theology of healing may be traced to earlier works of Friedrich Christoph Oetinger and even earlier to writings of Johann Albrecht Bengel.[15] But neither Oetinger nor Bengel put healing at the center of their theologies, as Christian Blumhardt did. Even more significant than Christian Blumhardt to the development of Christian healing practices was his nephew, Johann Christoph Blumhardt, who became the most important healing practitioner in Württemberg in the mid-nineteenth century.

Johann Christoph Blumhardt's healing ministry can be traced back to events that began in 1838.[16] During that summer, Johann Blumhardt took a pastorate in Moettlingen, where he noticed a total lack of enthusiasm among his parishioners. It was in his efforts to enliven his lethargic Moettlingen congregation, which often slept through his sermons, that Johann Blumhardt's healing ministry came to the fore. It began with the case of a demon-possessed parishioner named Gottlieben Dittus, whose deliverance took almost two full years, from 1841 to 1843.

In the early 1840s, Gottlieben Dittus experienced a variety of unusual symptoms such as convulsions and seizures, foaming at the mouth, unexplained bleedings, as well as visual and auditory hallucinations of spirits. Many of these symptoms, such as visions, hemorrhaging, and fainting spells, had been occurring since Dittus was a girl. Doctors could not explain many of these disorders, which more than once caused her to lose her job. Because of her pain and suffering, Dittus attempted to take her own life.

The first step Johann Blumhardt took in helping Dittus was diagnostic: exploring her past in order to determine what afflicting agents were responsible for her maladies. After thoroughly investigating her past, Johann Blumhardt learned that Dittus was abused by a guardian aunt who exposed her to popular peasant religion, which was widely practiced throughout the region. Johann Blumhardt determined that this exposure to peasant religion was directly responsible for Dittus' afflictions, which in turn were symptoms of demonic possession.

---

15. Oetinger believed that sin and demonic forces caused illness, while the role of demons in human suffering can be traced to Bengel (Macchia, *Spirituality and Social Liberation*, 26–27).

16. Earlier, while Blumhardt was working at the Basel Mission in the mid-1830s, there were instances of spirit possession and deliverance (Ising, *Johann Christoph Blumhardt*, 105).

In time, Johann Blumhardt ascertained that Dittus was possessed by over four hundred demons, which would often take control of her body and speak through her.

Diagnosis gave way to treatment. Johann Blumhardt, having no modern predecessor's methods to emulate, tried to follow Jesus' healing practices as recorded in the Bible. Over the course of two years, Johann Blumhardt prayed with Dittus night and day, fasted, and continually attempted to exorcise the demons. The final battle took place in late December 1843 when the demons also took possession of Dittus's brother and sister and threatened to kill them. Eventually, however, Johann Blumhardt drove the demons away and, as a result, Dittus made a full recovery, which included the disappearance of a high shoulder, a short leg, stomach troubles, and various other conditions.

Johann Blumhardt's theory of evil, derived from the exorcism of Gottlieben Dittus, made specific reference to aspects of popular peasant religion in Württemberg described as idolatry.[17] He defined idolatry as the belief in a (non-Christian) divine, supernatural, and invisible power by which a person could obtain health, honor, wealth, or other types of fortune. By the practice of idolatry, people became bound to satanic powers and were often possessed by demons. Demons could affect people physically by causing disease or by inducing antisocial feelings and behaviors such as lust, drunkenness, avarice, jealousy, and anger. Afflicting demons had to be forcefully cast out.

Along with forming Johann Blumhardt's theology of healing, Dittus's healing also sparked a revival in Württemberg. Hundreds of people flocked to Moettlingen to repent their sins, which often included participation in peasant religion. As a result, many were healed of various afflictions.

While the revival was taking place in January 1846, church authorities condemned Johann Blumhardt for praying for the sick instead of sending them to doctors. Physicians who put pressure on church authorities complained that Johann Blumhardt was infringing on

---

17. Popular peasant religion in Germany influenced the Grimm Brothers fairy tales, such as *Hansel and Gretel*, *Rumpelstiltskin*, and *Snow White*. Although from Hesse, not Württemberg, the Grimm brothers collected German folk tales that had variants from other German regions including Württemberg (Kamenetsky, *Brothers Grimm*, 118–23).

their (exclusive) right to treat the afflicted. Church authorities forbade Johann Blumhardt from praying for the sick and demanded that he refer them to physicians instead. While Johann Blumhardt never prohibited sufferers from seeking medical attention, he expressed distrust for biomedical means of healing, as he believed only God could heal the afflicted.[18] Johann Blumhardt believed Christians had to fight for their health with faith, repentance, and prayer. Eventually, however, Johann Blumhardt submitted to the church authorities and stopped praying publicly for the sick.

The revival in Moettlingen began to wane once Johann Blumhardt refrained from praying for the sick in 1846. Unhappy with this course of events, Johann Blumhardt decided to search for a new location to establish his healing ministry that would take him out of the church's sphere of authority. By 1852, Johann Blumhardt purchased a former health spa at Bad Boll, located near Moettlingen and Stuttgart, which became his *Kurhaus* or divine healing home.

Johann Bluhardt's *Kurhaus* was not completely independent from the Lutheran Church of Germany; Bad Boll was designated a "special" parish by the church. At Bad Boll, Johann Blumhardt was allowed to remain a pastor in the Lutheran Church of Germany, even though he received no salary from the church. He was permitted to carry out all pastoral functions, such as baptisms, communions, confirmations, weddings, and funerals, and the restriction against healing was lifted.[19] From 1852 until his death in 1880, Johann Blumhardt ministered to the sick and unfortunate at Bad Boll.

Johann Blumhardt's healing ministry was not an isolated phenomenon within Württemberg Pietism, but is an example of a much wider practice within southwest Germany in the nineteenth century. Another major figure from Johann Blumhardt's era, whose name is closely associated with his, is Dorothea Trudel. Trudel opened two healing homes in Mannedorf, Switzerland, in the canton of Zurich, between 1851 and 1856 to care for the sick. Located in German-speaking Switzerland, Zurich was part of the greater Württemberg religious community, as was the town of Basel. In the following section I will discuss areas in which Pietist healing and missions overlapped in greater Württemberg and in Ghana.

---

18. Macchia, *Spirituality and Social Liberation*, 75.
19. Zuendel, *One Man's Battle*, 102.

The Overlap of Healing and Missions in the Nineteenth Century

Missions and healing were both significant aspects of Pietist religious practice in nineteenth century Württemberg: Basel and Bad Boll were two of the four centers of Württemberg Pietism. Many leaders of the Peitist movement in Württemberg also participated in both spheres of church life.

The two most influential healing theologians, Christian Blumhardt and his nephew Johan Christoph Blumhardt, worked at the Basel Mission. Christian Blumhardt, who established the first systematic healing theology in Württemberg, was the Basel Mission's first Inspector (Director of Mission) after its establishment in 1815 until his retirement in 1838. Johann Blumhardt spent seven years (1830–37) at the Basel Mission teaching history, geography, Hebrew, Greek, Old Testemant and Old Testemant exegesis, Bible History, patristics, dogmatics, homiletics, catechetics, and the subject of "useful knowledge," which was a combination of mathematics, physics, and chemistry.[20] Upon leaving Basel in 1837 to work in Iptingen, Moettlingen, and finally Bad Boll, Johann Blumhardt remained closely attached to the Mission. Johann Blumhardt collected weekly offerings for the Basel Mission every Saturday evening at Bad Boll.[21] He was a frequent speaker at mission feasts, informed various congregations of the work of the mission, wrote a handbook about its history, and maintained close contact with missionaries throughout his life.[22]

One of the Basel missionaries who served in Ghana, Elias Schrenk, whose most notable accomplishments in Ghana included petitioning the British to keep the Gold Coast as a colony in 1865 and participating in early cocoa cultivation in the 1870s, was a strong proponent of Pietist healing.[23] Elias Schrenk entered the Basel Mission for training in 1854 at age sixteen and served as a missionary to Ghana from 1859 to 1872, where he first worked as general treasurer of the Basel Mission in Ghana.[24] In 1858, the year before he left for Ghana, Schrenk was ill and heard God tell him to visit Dorothea Trudel at Mannedorf to be healed. He first visited Johann Blumhardt at Bad Boll, and when he was not fully healed, he visited Trudel's home in August

---

20. Ising, *Johann Christoph Blumhardt*, 74–75.
21. Guest, *Pastor Blumhardt*, 52.
22. Meyer, *Translating the Devil*, 46.
23. Miller, *Missionary Zeal*, 19; Hill, *Migrant Cocoa-Farmers*, 171.
24. Danker, *Profit for the Lord*, 99–100.

1858. In March 1859, Trudel laid hands on Schrenk and healed him entirely, which allowed him to leave for Ghana.[25] In Ghana, Schrenk described being near death several times, only to be healed by Jesus at the last moment.[26] Schrenk is the most notable example of a Basel missionary strongly invested in Christian healing, although surely many other Württemberg missionaries in Ghana shared Schrenk's beliefs in Christian healing during the mid-nineteenth century.

These Basel missionaries never established a healing home like Bad Boll in Ghana. The Basel Mission did not systematically teach or practice Christian healing similar to the movement in Württemberg. One former missionary suggested that the Baselers could not tell the difference between the Holy Spirit and various demons in cases of possessed Ghanaians, which was a necessary criterion within the Württemberg healing system.[27] Another reason, frequently lamented by the missionaries in their reports, was that many Akan Christians did not recognize their sinful behaviors. Recognition of sin was another necessary criterion for the proper function of this Pietist healing system.[28] Whatever the reasons, the Basel missionaries never established systematic Christian healing practices based on the model of Bad Boll in Ghana.

There were, however, periodic accounts of healing miracles within the Basel community in nineteenth-century Ghana. The most notable healing occurred at Larteh in 1853, five miles south of Akropong, when (the Ghanaian) Catechist Edward Samson raised a boy from the dead with prayer, thereby precipitating a small revival. Thirteen baptisms resulted in the next six months in Larteh, and permission was obtained from the chief to open a school.[29]

Presbyterian leaders today, many of whom are involved in the healing and deliverance ministry, recall Basel missionaries as more

---

25. Boardman, *Record of the International Conference*, 152–53.

26. Ibid., 10, 125.

27. Meyer, *Translating the Devil*, 77.

28. Missionaries were constantly criticizing Akan converts for not recognizing their own (innate) sinful natures. In 1894 Rev. Rosler stated, "what is so painful in so many of our Christians is their scanty consciousness of sin. . . . Many in conversation claim to have no sin . . . and feel they are Christians when they have paid their church tax and been to church on Sundays." Evangel. Missiongesellschaft in Basel, The Ghana Archive, Series D-12, 1852–98, MF-5604 (hereafter EMB-GA), Annum Correspondence, Rosler's Report for the Year 1894, January 30, [18]95, no. II.155, reel 130, p. 310.

29. Smith, *Presbyterian Church*, 51.

committed to healing through prayer than the archival record suggests. A former moderator of the Presbyterian Church of Ghana from 1995 through 1999 wrote that the Basel missionaries in the nineteenth century healed the sick through preaching and prayer, and trained Ghanaian Christians to do the same.[30] Catechist Ebenezer Abboah-Offei, who currently manages the Presbyterian Church of Ghana's primary deliverance center, recalls his grandmother telling him that the Basel missionaries had the power of prayer, could perform miracles, and frequently healed people, especially in the Aburi hospital.[31] Similar sentiments were expressed by Edward Okyere, an Akropong Presbyterian, whose involvement in the parachurch organization Scripture Union in the 1970s helped to support the emerging Presbyterian healing and deliverance ministry.[32] Samuel Atiemo, pastor of the Ghanaian Presbyterian Reformed Church in Brooklyn, argues that the roots of the Presbyterian Church of Ghana came from Pietist Christianity, in which the European and Jamaican missionaries had the power of prayer and the ability to confront and destroy the works of the devil, which included oppressing Christians with disease and misfortune.[33]

From the historical accounts given by Ghanaian Presbyterian leaders today, one can deduce that certain missionaries and Akan Christian leaders did alleviate sickness and misfortune through prayer. But the Basel Mission in Ghana never established formalized healing practices, such as those found at Bad Boll. These healing practices, and the institution of the healing center, did in fact enter Ghana via the United States, after Württemberg healing practices became popularized in the late nineteenth century within the divine healing movement in the Anglophone world, particularly in America.[34] After the 1918–19 influenza pandemic, the Philadelphia-based Faith Tabernacle Congregation—a divine healing church—established a massive following in Ghana, particularly among Basel Christians, in the early to mid-1920s (see chapter 2).

But without well-established Christian healing practices, European and Akan Christians in nineteenth-century Ghana relied primarily

---

30. Beeko, *Trail Blazers*, 45.
31. Interview with Catechist Abboah-Offei, Akropong, Ghana, February 15, 2007.
32. Interview with Edward Okyere, Akropong, Ghana, February 11, 2007.
33. Lecture given at the deliverance workshop, Stony Point, New York, November 16, 2006.
34. Curtis, *Faith in the Great Physician*; Hardesty, *Faith Cure*.

on various forms of Akan therapy, which I will outline in the following section. Akan religion, like Württemberg Christianity, had a well-defined ideology about illness, health, and healing. Akan therapeutics, unlike that found in Württemberg, was heterogeneous, diverse, and inclusive, and its ministration involved a number of different ritual practitioners.

Akan Healing in the Nineteenth Century

Long before the Basel Mission began evangelizing in Ghana, forms of illness, misfortune, health, and healing that made reference to the supernatural were prevalent within Akan healing traditions. A reconstruction of Akan therapeutic traditions in the nineteenth and early twentieth centuries, during the time period predominately discussed in this chapter, is a complicated endeavor, because the first written accounts were composed by travelers, missionaries, and colonial anthropologists who, at times, were part and parcel of the institutions committed to destroying or controlling Akan healing and religious traditions.[35] These authors, however, were meticulous in their research, and by disregarding their moral judgments, one can reconstruct a partial picture of Akan therapeutic traditions. Because these are the very best and earliest written records of Akan therapeutics in English, I will fashion my reconstruction principally from these sources.

While multiple ethnic groups exist in Ghana, I will focus on the Akan for two reasons. First, they are the largest single ethnic group in Ghana; therefore, their healing traditions cover the widest area in the country.[36] Second, the Presbyterian Church was first established, and maintained its strongest presence, in the Akan regions of southern Ghana.

---

35. See Bowdich, *Mission from Cape Coast*; Christaller, *Dictionary*; Ellis, *Tshi-Speaking Peoples*; Rattray, *Ashanti*, and *Religion and Art*. Because therapeutics in Africa typically involves the supernatural, Western categories of religion and health significantly overlap in the African context. Thus, the history of healing in Africa closely resembles and overlaps with the history of religion (see Feierman and Janzen, "Introduction," 4).

36. This is not to suggest that Akan therapeutic practices were culturally or historically bound. Akan therapeutic systems, like those of other African ethnic groups, were extremely flexible and incorporative, especially compared with (the modernist discourse of) orthodox European medicine during the Victorian era.

In Akan religion, illness and misfortune often had a spiritual component, if not an ultimate cause, which needed to be treated through supernatural means. Health and prosperity for both the individual and the community were generally maintained through a series of rites of passage associated with birth, puberty, marriage, and death, which all involved elders pouring libations for the ancestors as well as worshipping family, clan, village, or national deities (*abosom*).[37] Annual rites such as the *Odwira* ceremony (yam festival)[38] and periodic rites such as the *Addae* festival (every forty days) were meant to purify and protect the town or nation from evil, with libations given to ancestors and other deities.[39] Ancestor and deity worship in the context of rites of passage amounted to protection against evil spirits—purveyors of sickness and misfortune—for both the individual and community.

In order to maintain the good health of individuals and the prosperity of the community, it was necessary to ensure that relationships among human beings, as well as relationships between humans and spiritual entities, were kept in balance through the reciprocal exchange of gifts. This included sacrifices to deities. If good health, peace, and prosperity were achieved, the rites described above were all that were performed. But if a person or group of people became ill or experienced some type of misfortune, help was sought from a variety of spiritual entities and their associated ritual practitioners.

An assortment of spiritual entities existed in nineteenth-century Akan religion that could have potentially assisted a sick individual or group of people, as in the case of a famine or epidemic. The sky god (*nyame*) did not significantly interact in people's lives owing to the more immediate influence of its subordinate deities (*abosom*).[40] There was no fixed pantheon of deities. Deities become more or less popular over time: some were disregarded, while others were introduced from different regions. Three major types of deities existed: regional deities (*ɔman bosom*), family deities (*abusua bosom*), and

---

37. For a more thorough discussion of Akan rites of passage, see Rattray, *Religion and Art*, 48–191. The *asamanfo* are defined as ancestral spirits (11), or spirits of the dead or ghosts (Christaller, *Dictionary*, 423).

38. For a significantly more nuanced analysis of *Odwira* in Asante, see McCaskie, *State and Society*, 144–42.

39. Rattray, *Religion and Art*, 127.

40. Christaller, *Dictionary*, 598.

prophetic deities (ɔkɔmfo bosom).⁴¹ These deities had anthropomorphic identities and were believed to be created by the sky god to execute his will regarding people. Each of the deities had different characteristics.

The regional deities were the tutelary spirits of a town, community, or state, while the family deities were the guardian spirits of a family or clan. Together, these two classes of spirits were called the old deities (*abosom-pan*) and inhabited places and things in nature such as rivers, hills, valleys, rocks, caves, trees, or forests.⁴² The old deities were served by priests (ɔsɔfo), who performed ritual duties to appease the guardian spirit. Some old deities were served by priests as well as prophets (ɔkɔmfo), who became possessed by the spirit, which spoke through them.⁴³ Sickness within the town, community, state, family, or clan was often interpreted as stemming from these old deities. This could occur if the followers of the deity neglected to perform proper rites for the deity or ignored a set of prohibitions for the deity.⁴⁴

In the case of an epidemic, the priests of the old deities would act. In 1883, for example, a whooping cough epidemic struck the southern Ghana town of Kukurantumi, and between sixty and eighty children died.⁴⁵ The reason for this calamity, argued the priests, was the rejection of the town deities by several Christian converts and many young men attending the Basel Mission school in the town. The adults asked the aid and protection of the town deities, which resided in a stone (*obo*). The priests predicted the destruction of the country by an imminent calamity if the townspeople did not seek protection from the town deity. At this point, the head priest told each family head to come to him with a cushion on his head to demonstrate his loyalty and also to swear an oath of allegiance to remain faithful servants of the deity.⁴⁶ The head of each family came to the house of the senior priest with a

---

41. Ibid.
42. Bowdich, *Mission from Cape Coast*, 262; Christaller, *Dictionary*, 598; Ellis, *Tshi-Speaking Peoples*, 12.
43. Christaller translates ɔkomfo as a prophet, soothsayer, or diviner (*Dictionary*, 598).
44. Bowdich, *Mission from Cape Coast*, 203; Rattray, *Religion and Art*, 3.
45. EMB-GA, Opoku's Report for the Year 1883, no. 74, January 3, 1884, reel 130, pp. 637–38.
46. The cushion on which head loads are carried in this context is the symbol of a slave, signifying subjugation. As slaves are subservient to their masters, these people, carrying cushions on their heads, are subjects of their deity (ibid.).

cushion on his head, at which time the town deity began to possess its prophets, speaking through them, and causing them to dance.

After this ritual, the priests began speaking publicly and going into individual homes to tell the inhabitants to offer sacrifices to the town deity in order to suppress its anger. The townspeople followed the priests' advice and made sacrifices of sheep and oiled mashed yams, and many children were dedicated for service to the town deity. The priests also put lit torches in front of the entrance to the Christian quarter, after which they immediately put them out with water, signifying that the impending calamity had been averted. In cases of large-scale suffering, such as this epidemic, the priests of the old deities were proactive.

The third type of major deity, the prophetic deities (ɔkɔmfo bosom), were referred to as the younger spirits (abosom-mma), and they had different characteristics than the older spirits, as they were focused on individual suffering. The prophetic deities were solicited by individuals in need, not by groups of people. The prophets of the younger deities were paid considerable amounts of money to assist in healing, to avert misfortune, to detect a person or witch who caused an illness, to expose a thief or adulterer, or to procure good luck.[47] Prophets (ɔkɔmfo) that served prophetic deities also dispensed medicines (aduru).[48] Prophetic deities did not reside in natural objects, but typically in various man-made objects and structures (see fig. 1.1). Another unique feature of the prophetic deities was that they were served exclusively by prophets, not by priests.

There were other spiritual practitioners besides prophets that were consulted by individuals to protect and heal, or alternatively, to inflict disease or misfortune on an enemy. These practitioners were medicine makers (aduruyefo), amulet makers (asumanfo), and herbalists (odunsini).

The medicine maker (aduruyefo) specialized in making medicines (aduru) to treat an array of personal afflictions as well as positively enhance a variety of social situations: sickness, war, wealth, pregnancy and birth, hunting and agriculture, success in business or at school,

---

47. Ellis, *Tshi-Speaking People*, 124. Christaller claims only the ɔkɔmfo bosom were consulted in sickness and misfortune (*Dictionary*, 598), while Ellis states that the abusua bosom were also consulted in sickness and misfortune (*Tshi-Speaking Peoples*, 92–93).

48. Christaller, *Dictionary*, 598.

Figure 1.1. "The fetish Odente [*abosom-mma*] in the Krobo farming district," 1899–1912. Ref. no. D-30.08.048. Photograph by Wilheim Erhardt. Reproduced with permission from the Basel Mission Archives / Basel Mission Holdings.

love affairs, and family conflicts.[49] Medicine was not always a consumable or applied object, but was often spiritual in nature (immaterial). Medicine was an ambivalent power, which could provide protection against evil, as in sickness, but also could be used to afflict an enemy.

An amulet or charm maker (*asumanfo*) made amulets (*suman*) that were generally used for personal ends and often kept in the user's house in order to cure sicknesses or alternatively to poison others.[50]

---

49. Rattray, *Religion and Art*, 40; Christaller, *Dictionary*, 101. The etymology of medicine maker comes from *aduru*, meaning medicine, drug, powder, or poison, and *ayefo*, which is a maker, author, mischief maker, or mischievous enemy (Christaller, *Dictionary*, 101, 587). Alternate spellings are *odu'yefo, oduyefo,* and *oduruyefo*. While the closest translation of *aduruyefo* is medicine maker, this type of Akan healer would most closely resemble the "sorcerer" from the African ethnographic literature (see Evans-Pritchard, "Sorcery").

50. Christaller, *Dictionary*, 600. Rattray refers to this practitioner as *Sumankwafo* (*Religion and Art*, 26). Christaller translated *asumanfo* as sorcerer, magician, or wizard (*Dictionary*, 483). Some evidence indicates that amulets were used to avert all evils

Amulets were endowed with spiritual powers that were distinct from the deities, although amulets were at times used in conjunction with deity supplication (see fig. 1.2).[51] The amulet's power came from the physical and spiritual properties of plants or trees used in its construction, as well as the spiritual power conferred on it from forest dwarfs/fairies (*mmoatia*),[52] forest monsters (*sasabonsom*),[53] witches (*ɔbayifo*),[54] or the spirits of the dead.[55]

Amulets and amulet making overlapped considerably with medicine and medicine making. Making and distributing amulets for various health and protective purposes was often referred to as making medicine.[56] Prophets (*ɔkɔmfo*) and medicine makers (*aduruyefo*) made amulets too.[57] Therefore, making amulets or making medicine was not the exclusive domain of either the amulet maker or the medicine maker. As this example indicates, the practice of any Akan healing practitioner frequently overlapped in any single individual.

A third type of Akan healing practitioner was the herbalist (*odunsinni*).[58] Etymologically, *odunsinni* is derived from the noun *dunsin*, meaning the stump of a tree.[59] The herbalist made medicine from roots as well as herbs, although there was still a spiritual component to this work.[60]

---

except sickness and death, which was dealt with in other ways (Bowdich, *Mission from Cape Coast*, 271).

51. Rattray, *Religion and Art*, 11–12. 23.

52. References to *mmotia* are found in Rattray, *Religion and Art*.

53. *Sasabonsam* is described as a forest monster with a huge human-shaped body, red skin, and long hair, who was believed to be the friend and chief of the sorcerers and witches (Christaller, *Dictionary*, 429).

54. The word ɔbayifo refers to the female witch, while *bonsam* refers to the male. Witches, almost always of the female variety, use their powers of destruction only within their own clan or family (Rattray, *Religion and Art*, 28). Witches eat people by sucking their blood and cause general misfortune to people within their families (29).

55. Ibid., 23.

56. Ibid., 26.

57. Christaller, *Dictionary*, 600.

58. The word *odunsinni* was translated as native physician, medicine man, charmer, sorcerer, or wizard (Christaller, *Dictionary*, 99). An alternate spelling is Dunsefo (Rattray, *Religion and Art*, 40).

59. Christaller, *Dictionary*, 99.

60. Rattray argues that the Akan herbalist (*dunseni*) or amulet maker (*sumankwafo*) believes that some leaf or root or plant is effective against a particular disorder because its spiritual potency is stronger than that of the affliction on which it acts (*Religion and Art*, 39, 20).

Figure 1.2. "Small patient with amulettes [*suman*]," 1881–1911. Ref. no. QD-32.008.0067. Photograph by Dr. Rudolf Fisch. Reproduced with permission from the Basel Mission Archives / Basel Mission Holdings.

Alternatively, material-based forms of healing such as inoculation and bone setting, which had little or no spiritual components, were also a part of the Akan healing repertoire in the nineteenth century.[61] Another form of nonspiritual medicine was called *abibiduru* ("native" or "country" medicine), which did not require a ritual specialist.[62]

Bone setting, inoculation, and native medicine aside, individuals experiencing various problems related to sickness and misfortune in the nineteenth and early twentieth centuries would consult these various practitioners to restore their health.[63] Afflicted individuals had several options: they could consult a prophetic deity, an amulet maker, an herbalist, or a medicine maker. Most of these Akan practitioners were morally ambiguous. Their powers to cure could also be used to afflict, depending on the wishes of the paying client.

While certain practitioners did compete over clientele, there was not a standard discourse of exclusion on par with that found in the Württemberg Pietist healing ministry, where participation in peasant religious healing practices was considered sinful and causative of afflictions. One Akan practitioner could accuse another of causing harm to a client. This was very logical, as practitioners were consulted at times to cause harm to other people. But types of practitioners did not try systematically to exclude other types of practitioners. The only spiritual entity referred to as systematically harmful were witches (ɔ*bayifo*).[64]

It was in this religious environment, which included these various therapeutic options for healing, that the Württemberg Pietists from the Basel Mission arrived to evangelize and spread their particular version of Christianity. These Akan healing institutions did not disappear with the establishment of Christianity through the nineteenth century, but instead coexisted, sometimes symbiotically and sometimes competitively, with churches. As I will show in the following section, through at least 1885, European missionaries and Akan Christians followed the local patterns of therapeutic inclusion established by Akan

---

61. See Maier, "Nineteenth-Century."
62. Christaller, *Dictionary*, 20.
63. There was another Akan practitioner, mentioned briefly in Christaller's *Dictionary*, but not in any of the other sources. This was the ɔsafo (not to be confused with ɔsofo, the priest), defined as one who cures a disease: curer, healer, physician (419). Perhaps this was a general term encompassing all Akan healers?
64. Christaller, *Dictionary*, 11.

non-Christians when they attempted to procure relief from illness and misfortune by consulting a variety of Akan healers.[65]

Basel Mission Therapeutics in Ghana, 1828–85

The physical separation of Akan Christians from non-Christians was the spatial embodiment of a discourse of separation from and a rejection of Akan religion. This physical separation was achieved by the Basel Mission through building a separate Christian quarter called Salem, based on the Korntal model, in every Ghanaian town where they had a church.[66] Korntal—another of the four centers of Württemberg Pietism—was a self-governing Pietist village kept separate from German society in order to maintain religious purity. Through separation, it was believed that moral Christian ethics could be better maintained in each Salem by Akan converts within the Christian quarter.

Along with the physical separation from Akan religion created by the Christian quarter, Basel missionaries offered Akan Christians means of spiritual separations from Akan religion. Spiritually, separation from Akan religion was marked by the ritual of baptism. At baptism the convert was made to renounce "the devil and all his works," which meant that the convert could no longer take part in any rites of passage or other rituals associated with Akan religion. Nor could an Akan Christian accept a stool office (chieftaincy), as political office was bound to certain Akan religious rituals, such as serving national deities, family deities, or ancestors. Throughout the Basel Mission yearly reports from 1850 onward, the tenor was one of antagonism toward "fetishes" (*abosom*/deity), "fetish-priests" (probably ɔkɔmfo/ prophet), and particularly the prophetic deities (ɔkomfo bosom), all of which were regarded as agents of the devil and the main obstacles to the spread of Christianity.[67] Peasant religion in Ghana, as it had been in Württemberg, was demonized. Demonizing and separating from Akan religion, both physically and spiritually, were predominant features of Basel Christianity.

---

65. In fact, research shows this was not a localized practice in Ghana, but within different regions throughout sub-Saharan Africa (Feierman and Janzen, *Social Basis*).
66. Jenkins, "Villagers as Missionaries," 430.
67. Smith, *Presbyterian Church*, 89; Middleton, "One Hundred and Fifty Years," 5.

But this separation was more of a discourse than a practice, particularly with respect to Akan therapeutics before 1885.[68] This discourse was produced particularly within Basel Mission reports. We should not take for granted that missionaries in Ghana were writing reports specifically for the Mission Board in Basel. This discourse of separation and demonization of Akan religion could have been produced to maintain forms of support from the Home Board. For example, Roman Catholic missionaries in East Africa would systematically misrepresent African religion and culture in the missionary press, embellishing African misery and savagery in order to increase their funding by patrons troubled by these images.[69] This might explain the contradiction between Basel missionaries demonizing Akan religion on the one hand, and consulting Akan healers on the other.

The spiritual separation between Akan religion and Christianity was neither maintained by Basel missionaries nor Akan Christians with regard to Akan therapeutics. In fact, several noted Akan Christians in the Basel community at this time were practicing Akan healers of several types, but particularly herbalists.[70] There was no spiritual healing alternative offered by the Basel missionaries; they did not introduce any form of Christian ritual protection against harmful evil forces, as was found at Bad Boll. In this environment, Akan healers saved the lives of missionaries and were frequently consulted by Akan Christians. While typically ineffective, missionary medicine became another choice within the many existing therapeutic options for both European and Akan Christians in Ghana during the nineteenth century.

---

68. There were several areas within the Basel mission project in Ghana where discourse and practices did not correspond. Miller (*Missionary Zeal*) refers to these disjunctures as organizational contradictions. Also, Hokkanen finds the same disjuncture within Scottish missionary circles in northern Malawi in the late nineteenth century (*Medicine and Scottish Missionaries*, 357–75).

69. Keiran, "Some Roman Catholic Missionary Attitudes."

70. See Fischer, *Der Missionsarzt*, 504–10. Other Akan Christians in the nineteenth century who practiced some form of Akan therapeutics included: Joshua Donko of Nsakye, who was an herbalist (*odunsinni*); Pastor Carl Reindorf, who was an herbalist; an unnamed Christian in Teshi, who by 1882 was treating Christians and non-Christians; and Catechist Noah Akwai, who was an Akan healer and successfully treated European missionaries in the 1870s.

In the following sections, I recount the cases of the European missionary Andreas Riis and the Ghanaian catechist N. Afwireng, who both suffered from severe illness, as evidence to support the above claims.[71]

The Case of Missionary Andreas Riis

In the nineteenth century, West Africa was referred to as the "white man's grave."[72] Travelers, traders, and missionaries succumbed to tropical diseases, particularly malaria and yellow fever, at very high rates. In response to this disease environment, the first group of Basel missionaries received medical training in Copenhagen before they left for Ghana, while later missionaries received medical training at the seminary.[73] At the seminary, Basel missionaries learned to treat tropical diseases with cold baths, bloodletting, and arsenic, which were techniques they used in Ghana.[74] These techniques, however, were typically ineffective and often harmful.[75] Therefore, the mortality rate—primarily resulting from malaria—was particularly high for Basel missionaries sent to Ghana in the nineteenth century.[76]

All four missionaries in the first group sent to Ghana died within three years after their arrival in 1828. Three of them passed away within the space of three weeks in August 1829, while the fourth, Johann Philipp Henke, died in November 1831. Henke died before the next three missionaries, Andreas Riis, Peter Petersen Jager, and Christian Frederich Heinze, arrived in March 1832. By the middle of July 1832, only Riis was still alive. In November 1837, two more missionaries arrived in Ghana from Basel, but by April 1938 both were dead. Between 1828 and 1838, nine missionaries were sent to Ghana and only Andreas Riis survived.

Like his colleagues, Riis suffered from a variety of ailments in Ghana, but he pursued different therapeutic options. Riis contracted

---

71. These two case studies were chosen because of the wealth of information about the specific illnesses in the Basel Mission reports. These two cases were not unique, but were typical among European missionaries and Akan Christians before the 1880s.

72. Fyfe, *A History of Sierra Leone*, 151.

73. Bartles, *Roots of Ghana Methodism*, 5.

74. EMB-GA, Wideman Extracts, Missions Magazine 1844, 187–92, reel 129, p. 72; no. 17, Joh Stanger. D. Accra, August 25, 1847, reel 129, p. 73.

75. Jenkins, "Scandal," 67; Fischer, *Der Missionsarzt*.

76. Fischer, *Der Missionsarzt*, 516.

malaria in September 1832 and was nursed at the plantation of George Lotterodt. Lutterodt was a European farmer who had been in Ghana since about 1805 and often cared for sick missionaries, such as Peter Peterson Jager, before Riis.[77] Riis recovered, not because of any treatment given by Lotterodt, but rather because of the care of an African doctor (*Neger-Doctor*) who prescribed cold ablutions and topical applications of lemon and soap. The healer who treated Riis was frequently consulted by the Christiansborg European community.[78] After this treatment, Riis fully recovered on the Akuapem hills, where he arrived on October 24.[79]

After his recovery in Akuapem, Riis accepted an invitation from Governor Helmuth V. Ahrensdorf to preach and teach European children within Christiansborg castle.[80] Riis lamented living and working among Europeans, as this appointment was contrary to his mission, but he argued that it was the best way for him to maintain good health.[81] But Riis's good health did not last long. By April 1834, Riis contracted another serious illness and was once again cured by a local healer.[82] During this illness, Riis claimed that Dr. Tietz, the European physician in Christiansborg, was useless. Riis argued that all patients treated by Dr. Tietz died, while those treated by the local healer survived.[83]

Riis was not the only missionary to be successfully treated by local Akan healers. Johannes Zimmerman, who wrote the first Ga dictionary and translated the first Ga Bible, was also cured of a serious illness by an Akan healer after biomedicine failed to restore his health. Both Riis and Zimmerman spoke of the transforming effect they experienced as a result of being cured by an Akan healer.[84] But who exactly were these local healers?

---

77. EMB-GA, no. 14, Riis 10, August 1832, p. 15.
78. Bartles, *Roots of Ghana Methodism*, 5.
79. EMB-GA, no. 15, A. Riis Ussue, December 2, 1832, reel 129, p. 15.
80. Bartles, *Roots of Ghana Methodism*, 6. At this time Riis was teaching the children of Commissary Richter, the high commissioner for war, as well as the daughter of the governor of Christiansborg (EMB-GA, no. 6, Riis to Buchelen, June 10, 1834, reel 129, p. 18).
81. EMB-GA, no. 6, Riis to Buchelen, June 10, 1834, p. 18.
82. EMB-GA, no. 5, Riis, April 1, 1834, p. 18. I cannot confirm whether the local healer Riis consulted during his second major bought of sickness was the same person or even the same type of healer. It is possible too that the healers consulted by the missionaries in and around Christiansborg were Ga healers and not Akan healers, as the Ga and not Akan comprise the local population in and around Accra.
83. Ibid.
84. Miller, *Missionary Zeal*, 144–45.

If one became sick in Ghana, there was a variety of Akan therapeutic practitioners from which to choose in order to potentially alleviate one's illness. The missionary reports usually referred to an "African doctor," "traditional doctor," or "fetish priest." There was the prophet (ɔkɔmfo), the amulet maker (asumanfo), the herbalist (odunsinni), and the medicine maker (aduruyefo), all of whom had the capacity to heal. All these healers made different sorts of medicine: some were material and consumed (like pharmacopeias), some were material and not consumed (like amulets), and others were spiritual and immaterial (like prayers). All four practitioners did incorporate some form of spiritual ritual into their healing techniques.[85] Healers of any one of these types, or some combination thereof, could have treated Riis, Zimmerman, and many other Basel missionaries. These missionaries were following an Akan therapeutic pattern that was inclusive, unlike the exclusive forms of religious healing popularized at Bad Boll. Akan Christians, like the catechist N. Afwireng, took a similar course of action when ill.

The Case of Catechist N. Afwireng

Akan Christians, like European missionaries, suffered from a variety of afflictions in nineteenth-century Ghana. An 1887 missionary report from the Kingdom of Kwahu (located northwest of Akuapem) indicated that smallpox, measles, and whooping cough were endemic. Most people had skin infections, and a significant number suffered from leprosy, fever, and dysentery as well.[86] Because of the high frequency of the sickle-cell gene in Akan populations, malaria was not as statistically fatal for local populations as it was for European missionaries.

For most of these diseases, missionary medicine was completely ineffective. Akan Christians, like the non-Christian Akan and the European missionaries in the nineteenth century, followed a therapeutically inclusive model, in which a number of different healers and types of healer were consulted to find relief from various

---

85. All the reports referred to a therapeutic practitioner, therefore I conclude that the Christians were not treated with country medicine (abibiduru).

86. EMB-GA, Schmid's Report to Basel on Poverty, Illness, and Accident in the Social Life of Kwahu, no. KK.120, October 10, 1887, reel 129, p. 240.

afflictions. One particularly detailed account followed the catechist Afwireng, whose illness and treatment in 1885 was recorded in a Basel Mission report.[87]

In early 1885 the catechist N. Afwireng, who worked at the Basel station in Abetifi, traveled to the town of Obo and, afterward, became chronically ill. In two months he developed severe pains in his arms and legs, and was near death. Afwireng was taken to Akropong, where he arrived with a swollen foot. Reverend Dilger, a Basel missionary, believed that the swollen foot was a symptom of his previous illness and prescribed rest in order to cure this affliction. "Traditional doctors" were consulted, who told Afwireng he was being spiritually poisoned. This convinced Afwireng of the supernatural root of his illness.

The catechist needed another expert opinion; therefore, Afwireng consulted another Akan healer named Kwabi (see fig. 1.3)—who at this time, if not later, became a catechist (and possibly a pastor) in the Basel Mission.[88] A passage describes Kwabi's treatment of Afwireng as "drastic" without being more specific. Kwabi also prescribed Afwireng a substance to rub into his skin and a therapeutic liquid to drink. A few days later, Afwireng began to have fainting spells, which caused him anxiety. Many, including Afwireng himself, believed his life was in danger. Kwabi commented that his good medicine would be of no use if God had already decided to take his life. This statement caused Afwireng extreme anxiety. As the fainting spells continued, the missionaries suggested that Afwireng travel with Kwabi to his home in Mpraeso to receive treatment for an extended period of time.

Before traveling to Mpraeso with Kwabi, Afwireng asked for Dilger's thoughts about consulting another Akan healer, whom Dilger described pejoratively as "not only a heathen but [also] a bad one." This healer, whose name was not given, was a relative of Afwireng and was eventually consulted by the ailing catechist. This new healer deduced that the "fetish priest" (probably an ɔkɔmfo/prophet) from Obo had spiritually poisoned Afwireng. As a result, the healer advised Afwireng and Kwabi to claim the money they lost (presumably from

---

87. The account of Afwireng's illness is in: EMB-GA, Dilger to Basel, May 20, 1885, no. II.124. reel 129, pp. 199–202.

88. Fischer, *Der Missionsarzt*, 504–5. Figure 1.3—a photograph from the Basel Mission Archive—lists Kwabi as a pastor, not a catechist.

Figure 1.3. "Pastor Kwabi with three Africans," 1885–1911. Ref. no. D-30.11.019. Photograph by Dr. Rudolf Fisch. Reproduced with permission from the Basel Mission Archives / Basel Mission Holdings.

lost work and therapeutic costs) since the beginning of the illness from the Obo prophet.[89]

Both healers—Kwabi and Afwireng's relative—told the Obo prophet that he should come to Akropong and heal Afwireng. The prophet denied that he poisoned Afwireng and swore the Kwahuhene's oath (oath of the chief of Kwahu) that it was not he who caused Afwireng's illness.[90] The case was brought before the Kwahuhene's court, which ruled against the Obo prophet. Kwabi was awarded twenty dollars in order to cover the cost of Afwireng's treatment and to compensate him for lost time. Reverend Dilger intervened at this point and told Kwabi it was inappropriate to accept this

---

89. The passage claims that since this healer was a relative of Afwireng, he could not claim money for his services. Rev. Dilger, who wrote this report, believed this healer initiated a lawsuit in order to procure funds.

90. Oaths taken in the context of a legal dispute were meant to attest to the truthfulness of the statement (Ellis, *Tshi-Speaking Peoples*, 198). Obo, like Abetifi, is a town in the Kwahu kingdom, therefore the oath taken would have been that of the local political authority, the Kwahuhene, or chief of Kwahu.

money.[91] Instead, the Basel Committee promised to reimburse Kwabi twenty dollars for previously treating Afwireng.

While the lawsuit was settled and Kwabi was compensated in principle, Afwireng was still very ill. The missionaries suggested that Afwireng be sent to Dr. Rudolf Fisch, the medical missionary in Aburi who had arrived in Ghana earlier in 1885, as a last attempt to cure him. After Afwireng's arrival in Aburi, Deacon Obeng, who was Afwireng's brother, requested that Afwireng be sent to him in Nsakye. This request was apparently denied, and Afwireng was examined and treated by Fisch. Fisch observed that Afwireng had multiple health problems: anemia, damage to his left lung, an enlarged spleen, abnormally numerous white blood corpuscles of the large linear type, as well as eye problems, urine problems, and a psychological condition due to loss of hope. Fisch's final diagnosis was lineal anemia and hysterical melancholy. For this, Fisch prescribed arsenic in mounting doses. Afwireng only stayed four days at the hospital; the account of his illness ends when he left Fisch's care.

During the course of his illness, Afwireng consulted up to twenty different Akan healers as well as the missionary doctor for treatment of his illness. One missionary reported that the Akan healers consulted by Afwireng were mostly "fetish priests" and "soothsayers" (probably ɔkɔmfo).[92] One healer made a "fetish" (suman/amulet) for Afwireng in order to treat his illness: this person may have been an amulet maker (asumanfo), prophet (ɔkɔmfo), or medicine maker (aduruyefo). All parties in Ghana—European missionaries, Akan Christians, and non-Christian Akan—relied on Akan healers primarily for their healthcare until at least 1885.

Therapeutic Shifts in Württemberg and the Missionfield, 1885–1918

Toward the end of the nineteenth century, Württemberg Pietist therapeutics began to shift from Christian healing, of the sort institutionalized at Bad Boll, to biomedicine. This shift, driven by the bacteriological revolution in Europe, affected the Basel Mission, which

---

91. Dilger felt it was immoral to "take vengeance in this way," meaning, perhaps, seeking retribution from a prophet.

92. BMG-GA, Schmid's Report to Basel on Poverty, Illness, and Accident in the Social Life of Kwahu, 240.

in 1885 sent its first medical missionary, Dr. Rudolf Fisch, to Ghana. From the 1880s onward, a gradual shift occurred among Basel missionaries in Ghana away from consulting Akan healers and toward relying exclusively on biomedicine. This caused a significant problem for Akan Christians. Many believed that illness and misfortune had supernatural causes and thus needed to be redressed by supernatural means. Consequently, many Akan Christians continued to rely on Akan healers for a variety of afflictions.

Shifts in Württemberg Healing: From Spiritual Therapy to Biomedicine

In his later years at Bad Boll, Johann Christoph Blumhardt's theology shifted. In the early part of Blumhardt's ministry, he viewed healing as a purely supernatural process enacted by the Holy Spirit and accomplished by invoking the name of Jesus. He strongly questioned the effectiveness of biomedical means of healing. The tension between his healing system and the medical establishment was exacerbated when the medical community, which lobbied church authorities, succeeded in forcing Blumhardt to stop performing Christian healing rituals, such as laying on of hands and praying for the afflicted. But later in Blumhardt's life, he became disillusioned with the effectiveness of healing through prayer and was convinced that medicine played a significant role in healing, even to the point of recommending that most of the sick that came to Bad Boll should consult physicians.[93] Toward the end of his life, before his death in 1880, Johann Blumhardt believed that doctors served as instruments through whom God provided healing; this was a discourse adopted by medical missionaries generally during this time.[94]

The bacteriological revolution in Europe significantly influenced this new investment in biomedicine by Johann Blumhardt and within Württemberg Pietism during the 1880s. During this time, a series of discoveries of specific causal agents of infectious disease transformed medical practice and theory in Europe. These scientific advances significantly influenced evangelical missions and contributed to the emergence and development of medical missions generally.[95] By the

---

93. Macchia, *Spirituality and Social Liberation*, 75.
94. Ranger, "Godly Medicine."
95. Grundman, *Sent to Heal!*, 45–51.

time of Blumhardt's death in 1880, healing evangelism and practices had begun to wane in Württemberg, as evidenced by the theological shift of Blumhardt. This change, from Christian healing to biomedicine within late Württemberg Pietism, also corresponded with a shift in practices in the mission field.

Shifts in Mission Therapy: From Akan Healers to Medical Missionaries

Before 1885, the Basel Mission and Basel missionaries in Ghana had little investment in biomedicine, although instruction in personal healthcare in the tropics was formally taught in the seminary to outgoing missionaries after 1845.[96] The first Basel Mission doctor, C. F. Heinze, arrived more than fifty years prior to Dr. Fisch in 1885, but Heinze died six weeks after his arrival in 1832. After Heinze's death and up to 1885, several missionaries requested that physicians be sent to Ghana. By 1843, Rev. Widmann suggested to the Home Board that it would be good to have a Christian doctor in Ghana.[97] In 1863, Elias Schrenk expressed plans to the Home Board for the construction of a mission hospital with an accompanying physician.[98] In 1865, the head of the (British) Government Medical Office, who had treated the Basel missionaries for a long period, decided to leave and suggested that the Mission send its own doctor to Ghana. A year later, in 1866, J. G. Christaller's wife died, after which he asked the Mission to send a doctor to study local health practices, medicines, and illnesses.[99] In 1870, Elias Schrenk's third request for a missionary doctor was given more consideration by the Mission.[100]

Finally, in 1882, Dr. Ernst Mahly was sent to Ghana with the task of investigating, over a two-year period, the particular health problems there. Mahly's report gave several recommendations with regard to the prevention of disease through proper clothing and diet, as well as

---

96. Fischer, *Der Missionsarzt*, 110.
97. BMG-GA, Wideman Extracts, Missions Magazine 1844, 187–92.
98. This project was unable to continue because of a lack of funding, and the hospital was abandoned in 1866 (Fischer, *Der Missionsartz*, 139–40).
99. Ibid., 102–3.
100. Ibid. Schrenk was also troubled that Ghanaian Christians, when sick, would consult with Akan prophets ("fetish priests"), which in his opinion was regressing to their previous religion. Here you see therapeutic choices being judged as theological ones.

the treatment of common diseases. As a result of this report, the Basel Home Board decided to send medical doctors and nurses to Ghana. Some of the board members, unhappy with this move, considered the decision to send doctors and nurses indicative of a lack of faith.[101] This faction of the Home Board, upset with sending medical missionaries, believed that Christians should trust in God alone for their healing, in agreement with Johann Blumhardt's earlier approach.

When Fisch was appointed to Ghana as a medical missionary in 1885, he established a sanitorium and dispensary in Aburi, a town on the Akuapem ridge near Akropong (see fig. 1.4). Fisch's responsibilities included caring for the missionaries at scattered stations and conducting outpatient clinics at the two medical stations in Aburi and Abokobi.[102] The small hospitals at Aburi and Abokobi grew steadily, owing to an increase in treatment of both Ghanaians and Europeans. In 1912, there were over 20,000 outpatients treated at Aburi and Abokobi; 129 were admitted to the twelve hospital beds.[103] The Home Board decided to build a modern, well-equipped hospital at Aburi, but the project was halted because of the deportation of the Basel missionaries in 1918.

As biomedical means of healing became more effective in the first decade of the twentieth century, there was much less reliance by missionaries in Ghana on Akan healers. Missionaries like Riis and Zimmerman owed their lives to Akan healers. But by the time of Catechist Afwireng's illness in 1885, some of the missionaries were skeptical of the idea of being spiritually poisoned by malevolent spiritual forces and hence questioned the effectiveness of Akan healers.[104] This attitude, of course, was enabled by the treatment of malaria with quinine that came into practice by the mid-nineteenth century.[105] During Fisch's tenure in Ghana, he was able to reduce missionary

---

101. Smith, *Presbyterian Church*, 186.

102. However, Fisch's primary medical duty, as instructed by the Mission, was the care of the European missionaries (Fischer, *Der Missionsartz*, 201).

103. Smith, *Presbyterian Church*, 187. A year earlier (1911) the Basel Mission established a public health board (Brokensha, *Social Change*, 22).

104. Subscript to the letter: BMG-GA, Ramseyer, Dilger to Basel, May 20, 1885, 200. Alternatively, one missionary named Edmund Perregaux, who was stationed in Ghana from 1891 to 1905, did report cases where he was convinced of occult powers afflicting Akan Christians in Ghana (Fischer, *Der Missionsartz*, 501).

105. Fischer, *Der Missionsartz*, 516. By 1875, the British introduced quinine as a malaria prophylactic in Ghana (Curtin, *Migration and Mortality*, 88).

Figure 1.4. "Dr. Fisch's dispensary, Aburi, at the turn of the century," 1897. Ref. no. QW-30.006.001. Photograph by Dr. Rudolf Fisch. Reproduced with permission from the Basel Mission Archives / Basel Mission Holdings.

mortality from 36 percent to 6 percent through the prophylactic use of quinine.[106] As biomedicine became more of an empirical and rationalist science, it shut itself off from non-Western forms of knowledge, thereby derogating their value. African healing systems were reduced to "magic" and "herbs" by Europeans.[107]

After Fisch's arrival, the therapeutic options supported by the missionaries shifted toward biomedicine, but not immediately.[108] While 1885 marked the beginning of a transitional period, evidence suggests that some Basel missionaries continued consulting Akan healers into the twentieth century.[109] For Akan Christians, biomedicine did not address the evil spirits believed to be the cause of many illnesses and cases of misfortune, nor did the Basel Missionaries acknowledge

---

106. Fischer, *Der Missionsartz*, 518.
107. Comaroff and Comaroff, *Revelation and Revolution*, 328.
108. See Fischer, *Der Missionsartz*, 519.
109. Jenkins, "The Scandal," 68; Fischer, *Der Missionsartz*, 520.

the existence of these malevolent spirits.[110] Akan Christians, alternatively, did believe in the ability of harmful spirits to afflict them and consulted Akan healers in the latter years of the nineteenth century and into the twentieth century. This caused considerable friction with European missionaries, who more and more complained of the "superstitious Africans." At this time, therapeutic choices by Akan Christians began to be judged more as theological choices by the Basel missionaries.[111]

In a report from 1894, Rev. Rusler described a problem within the Basel Christian community in the town of Anum, where many local Christians still observed superstitions and feared evil spirits.[112] An elder in the church believed that there was a wicked spirit in his house causing him misfortune and anxiety. Rev. Rusler lamented that the elder was not content to pray about it but secretly went to an "Osumanni" (amulet maker/*asumanfo*), who gave him an amulet as well as medicine to place around his house. Rev. Rusler concluded that the missionaries must continue to fight against these practices that were contrary to God's word. When confronted by missionaries, local Christians in Anum tried to explain these forms of therapy as local customs that should be treated by their European brethren with a certain amount of cultural relativity.

Conclusion: Therapeutic Tensions in Ghana's Basel Mission

Through the first ninety years of the Basel Mission in Ghana, from 1828 to 1918, the missionaries tried to replicate a Württemberg Christian peasant society in which agriculture, husbandry, and crafts were promoted, as well as the physical and spiritual separation from the deleterious influences of non-Christian peasant society. A strong discourse emerged among the missionaries portraying Akan religion as demonic.

---

110. Fischer, *Der Missionsartz*, 502.
111. For a parallel development among the Anglican community in Tanzania, see Ranger "Godly Medicine."
112. BMA-GA, Rusler's Report for the Year 1894, January 30, 1895, no. II.155, reel 130, p. 309. The people of Anum are ethnically Guan, not Akan. Because of the long history of interaction between Guan and Akan peoples (particularly in Akuapem), I will assume that these occurrences in the Basel Mission Guan community of Anum would have been the same as in Basel Mission Akan communities.

But this discourse did not correspond with the Basel Christian community's participation in Akan therapeutics through the mid-1880s.

The Württemberg missionaries came from a society where Christian spiritual healing practices were prevalent, but were not able to replicate these religious healing institutions in Ghana. During this period, from 1828 to around 1885, the Basel Mission was partially enchanted. Missionaries like Elias Schrenk, who came out of the tradition established by Johann Christoph Blumhardt, explained suffering in the world in spiritual terms, which corresponded to a belief in spiritual entities that could alternatively heal and afflict. The Basel Mission, however, did not supply or train practitioners to manage the therapeutic process in Ghana. Instead, the mission informally outsourced these spiritual healing practices to Akan healers (while simultaneously and ironically demonizing them).

From 1885 through the early twentieth century, a shift gradually occurred, to an extent driven by the scientific discoveries of the bacteriological revolution. Biomedical forms of healing became more accepted and eventually were promoted exclusively by missionaries in Ghana as well as in Württemberg. Missionaries, who began questioning the validity and legitimacy of Akan therapeutic practices, consulted Akan healers less frequently. One unintended consequence of this transition was increasing patriarchy within the Basel Mission, because it was very common for women to act as Akan healers, particularly as prophets, while women rarely acted as religious leaders within the mission (see chapter 6 for a fuller discussion of gender and Christian therapy).

This transition away from incorporating the practices of Akan healers, in turn, disenchanted the Basel Mission: illness was considered material and treatable via biomedicine. While the European missionaries began to use biomedicine exclusively for treatment of their diseases, Akan Christians continued to solicit help from Akan healers. At times this solicitation occurred covertly, as European missionaries discouraged this practice along with colonial authorities, who frequently outlawed various forms of Akan healing.[113] Only Akan healers, however, provided protection from malevolent spiritual forces believed to be the cause of many diseases and disorders. This void—between disenchanted missionary discourse and enchanted Akan Christian practice—was addressed more systematically and overtly by many Akan Christians beginning in 1918.

---

113. See for example Gray, "Witches, Oracles, and Colonial Law."

2

Enchanted Competition for the Presbyterian Church of Ghana, 1918–60s

Between 1918 and 1960 the Presbyterian Church of Ghana was disenchanted. The church denied the existence of many afflicting spiritual forces, such as witches, even while witchcraft accusations flourished in Akan society, particularly during the cocoa boom. Correspondingly, the Presbyterian Church of Ghana did not offer methods of religious healing or protection from these spiritual afflictions. The church considered Akan healers, who could treat these spiritual disorders, illegitimate; no longer were they used as an outsourced form of therapy by the Christian community as they had been before 1885. In fact, church members who consulted Akan healers were often excommunicated. The threat of excommunication, however, did not deter members of the Presbyterian Church of Ghana, who frequently sought Akan healers covertly to manage their illnesses and misfortunes. After 1918 Basel Christians had more options for spiritual therapy, including within Christianity. This year marked the beginning of enchanted Christianity in Ghana, which seriously challenged the Presbyterian Church of Ghana, particularly in regions where the church was most influential.

Cocoa, Migration, and Witchcraft

The development of the cocoa industry in West Africa to a significant extent depended on noneconomic institutions and relationships as well as market mechanisms.[1] In Ghana specifically, the Basel Mission played an important role in the early development of the cocoa industry, particularly in the Akuapem state, whose capital Akropong

---

1. Berry, *Cocoa, Custom, and Socio-Economic Change*, 8.

was the headquarters of the Basel Mission. By the mid-1860s, cocoa seeds were planted at Basel Mission stations at Aburi, Mampong, and Krobo-Odumase.[2] Cocoa growing progressed in Mampong and Odumase, and by the 1870s cocoa growing experiments were undertaken in Aburi by Rev. Elias Schrenk. The Basel Mission also introduced cocoa directly into Akyem Abuakwa in 1890, when Rev. Adolf Mohr (no relation to author), who headed the Basel Mission station in Begaro, began distributing cocoa pods among Akyem Christians.[3]

The popular account of cocoa expansion in Ghana focuses on Tetteh Quashie, a Ga blacksmith from Christiansborg, who established a cocoa nursery in Mampong (Akuapem) with pods brought back from Fernando Po in 1879. Alternatively, Rev. Mohr possibly sold cocoa pods to Tetteh Quashie.[4] Regardless of how Quashie acquired the pods, he sold and distributed cocoa pods and seedlings to various Akuapem farmers. After Quashie's success, the colonial government took part in and strongly encouraged cocoa cultivation, and by 1895 young cocoa trees were thriving in the Government Botanical Station at Aburi (Akuapem).[5] In 1896 or 1897, pioneering farmers from several Akuapem towns started buying forestland for cocoa plantations in various parts of Akyem Abuakwa, located west of Akuapem across the Densu River, initiating the large-scale cocoa industry.

The early cocoa industry was integrally tied to the Basel Mission in a number of other ways besides the early cocoa growing experiments. The first cultivators of cocoa in Akyem Abuakwa by the 1890s were Akuapem men, many of whom were ministers, catechists, teachers, and prominent Christians in the Basel Mission.[6] The best plantations before 1900 were maintained by two Basel ministers, one Jamaican (Rev. A. W. Clerke) and the other Ghanaian (Rev. Sampson). Likewise, the Basel Mission Trading Company quickly built up

---

2. Hill, *Migrant Cocoa-Farmers*, 171.
3. Ibid.
4. Ibid., 171–72.
5. The establishment of the Aburi gardens in 1886 was primarily to teach Ghanaians how to cultivate crops in a systematic manner for the purposes of export, particularly cocoa seeds, which Governor William Griffith had imported from São Tomé. Griffith strongly encouraged cocoa growing in public meetings with chiefs and villagers in Akuapem in the 1890s (Hill, *Migrant Cocoa-Farmers*, 174; Debrunner, *History of Christianity*, 253).
6. Hill, *Migrant Cocoa-Farmers*, 168. Those who were Basel-schooled or literate were also in high demand as secretaries or bookkeepers for cocoa growers (Brokensha, *Social Change*, 17).

a system of purchasing stations and of transporting harvested cocoa to the coast.[7] In addition, by 1898 European exporting firms, the earliest being the German West African Trading Company, agreed to handle the cocoa if transported to the coastal ports, an arrangement facilitated by the Basel missionaries.[8] The Basel Mission Trading Company, in fact, exported the first commercial shipment of cocoa to Hamburg in 1893.[9] Through Basel Mission networks, cocoa was grown, transported, and shipped to Europe.

Cocoa production, however, quickly expanded beyond the Basel Mission network. Cultivation was taken up in nearly every district of southern Ghana, but by 1910 the Eastern Region produced about 75 percent of the cocoa exports.[10] Nearly all the production was done by local Ghanaian planters, who engineered the rapid growth of cocoa production on distinct capitalistic lines.[11] Ghanaian cocoa farmers frequently formed companies to raise money to purchase land and used money earned from initial farms to reinvest in new cocoa farms.[12] By 1911, Ghana became the world's largest cocoa producer, and by 1920 cocoa amounted to 83 percent of Ghana's export earnings.[13]

One particular feature of this emerging industry in Ghana was the urban to rural migration of Ghanaian cocoa farmers. Cocoa farming was usually done at a distance from one's hometown. In the Eastern Region, migrants originally from Akuapem towns, and later from other towns such as Anum and Boso, bought property in forested areas, where they established a farm and settled nearby, creating new villages. Cocoa farms in Asante, however, were farmed by Asante predominantly, not outsiders, and the farms were often located closer to the farmers'

---

7. Debrunner, *History of Christianity*, 253.
8. Ibid.
9. Southall, *Cadbury on the Gold Coast*, 26.
10. Hill, *Gold Coast Cocoa Farmer*, 106.
11. Ibid.
12. The group was organized solely for the purchase of the land, not for joint or communal ownership. Once lands were purchased, individuals would get a portion of the total land proportionate to the money contributed to the company for the purchase of the land (Hill, *Migrant Cocoa-Farmers*, 38–39).
13. Kay, *Political Economy*, 15. Not only cocoa growers, but also cocoa transporters made significant profits: transporters in 1903 could make more money taking loads of cocoa to the coast in nine days than they could working for the Basel Mission for an entire month (Debrunner, *History of Christianity*, 254).

Figure 2.1. "Opening the cocoa pods. Families work together on their own farms." Family processing cocoa pods on a farm in southern Ghana during the 1920s. Notice the cocoa broker or purchaser wearing a pith helmet in the rear. Maxwell, *Gold Coast*, 187.

homes. Cocoa farms in Asante were therefore located originally on the outskirts of bigger cities such as Kumasi and Sunyani.[14]

This large-scale labor migration significantly depopulated previously established towns, particularly in Akuapem, which concerned the Basel missionaries. Between 1911 and 1921 the population in Akropong dropped from 6,281 to 1,226, while during this same period the population of Nsawam, a cocoa town and principle destination for Akropong farmers, increased from 2,569 to 6,143.[15] The Basel Mission tried to adjust to the large-scale migration of many of its members, who often left for weeks and months at a time, by building outstations in Eastern Region cocoa towns such as Mangoase, Suhum, Nsawam, and Asamankese.[16] In particular, a Presbyterian

---

14. Mikell, *Cocoa and Chaos*, 72. Outsiders did not migrate to Asante until the 1940s, when cocoa farmers from the Eastern Region migrated into undeveloped parts of Asante (Berry, *No Condition is Permanent*, 73).

15. Hill, *Migrant Cocoa-Farmers*, 248.

16. Debrunner, *History of Christianity*, 253.

congregation was established adjacent to Adawso, the commercial center of the cocoa trade, in Apasare in 1894.[17]

While the Basel Mission helped to facilitate the early cocoa industry in Ghana, by the turn of the twentieth century the missionaries were frustrated by the accompanying social changes of the large-scale migration. One missionary wrote in 1907 that "Cocoa is spoiling everything . . . the Akuapem stations present a disagreeable picture, most of the people are away on their cocoa farms . . . they seldom attend Church and are most liable to influence from pagans . . . everywhere there is discontent, quarrels, irregular living, open strife."[18] Individual wealth resulted from cocoa, but so did social chaos.

The cocoa industry instigated rapid social change by altering the lineage-based subsistence economy. Cocoa production strengthened individual male farmers but weakened matrilineal families.[19] Money earned from cocoa farming was considered the property of individual farmers, not the matrilineage, as was the case with other sorts of earnings. Furthermore, many farmers were able to bypass the traditional inheritance system and bequeath their cocoa farms to their sons, not to their matrilineage.[20] This caused intense conflict with respect to the distribution of wealth and inherited property between the conjugal family and the matrilineage.

Besides transforming the lineage-based subsistence economy, cocoa farming strongly contributed to patriarchy in Ghana during the colonial period. Women played central roles in securing loans as pawns for male kin who needed capital to buy land for cocoa farming. And while women accounted for a significant amount of the productive forces within the cocoa economy, they rarely owned their own cocoa farms or reaped financial benefits from working on their male relatives' or husbands' farms. The cocoa economy significantly increased the hardships for women in Ghana.[21]

Migration, wealth, and structural changes in social relationships within the cocoa industry created significant social strife for people, many of them Basel Christians or Basel-schooled.[22] Many people

---

17. Hill, *Migrant Cocoa-Farmers*, 219.
18. Smith, *Presbyterian Church*, 138.
19. Allman and Tashjian, *"I Will Not Eat Stone."*
20. Hill, "Migrant Cocoa Farmers," 216.
21. Allman and Tashjian, *"I Will Not Eat Stone."*
22. Cocoa farming also negatively affected one's physical health in certain ways. Clearing forests for cocoa farms created ideal conditions for *A. gambiae*, a significant

who lived through that period remarked that "cocoa destroys kinship, and divides blood relations."[23] This widespread social unrest within families provoked accusations of witchcraft, which was believed to be affective and inherited within the matrilineage, representing the "dark side of kinship."[24] During this period, witchcraft was believed to have the power to burn down a cocoa farm or cause poor crops, in addition to other woes such as sterility, impotence, and various other diseases and misfortunes.[25] Witchcraft was particularly known for killing children.[26] More generally, witchcraft in Ghana was symptomatic of a conflicted and unstable society.[27]

The formal stance of the Basel Mission throughout this period was to deny the existence of witches and to excommunicate Christians who accused anyone of witchcraft.[28] Witchcraft confessions and accusations were not divorced from Presbyterian institutions, however. As late as the 1950s, for example, students of the Presbyterian Girl's School and Women's Training College at Agogo declared themselves witches and terrorized the school.[29] From the 1920s through the 1950s, public opinion rendered the Basel Mission/Presbyterian Church helpless against witchcraft attacks. The church focused its efforts on biomedical means of healing: building hospitals, clinics, and supporting public health measures.[30]

The explosion of social dislocation and witchcraft during the cocoa boom demanded extraordinary redressive measures, and new forms

---

malaria vector. Resources devoted to cocoa were diverted from food crops, and the nutritional consequences of decreased food production did affect more marginalized people (Patterson and Hartwig, "Disease Factor," 13–14).

23. Allman and Tashjian, "*I Will Not Eat Stone,*" 123; Mikell, *Cocoa and Chaos,* 112.
24. Geschiere, *Modernity of Witchcraft,* 11.
25. Debrunner, *Witchcraft in Ghana,* 42–44.
26. The etymology of *bayi* possibly derives from *oba* (child) and *yi* (to remove) (Christaller, *Dictionary,* 11).
27. Allman and Parker, *Tongnaab,* 134.
28. Debrunner, *Witchcraft in Ghana,* 135–37.
29. Ibid., 3.
30. In 1931, the Presbyterian Church of Ghana at Agogo opened a hospital, while dispensaries were opened in nearby towns. In 1935, provisions were made to treat lepers, and in 1951 the church staffed a clinic in Dormaa Ahenkro, and in Bechem the following year, after which the church accepted the task of running a government hospital in Bawku (Beeko, *Trail Blazers,* 45). In Larteh, the Basel Mission was responsible for creating a public health board in 1911 (Brokensha, *Social Change,* 22). The Basel/Presbyterian community probably played a role in promoting public health measures in other towns too.

of therapy were introduced in Ghana, such as antiwitchcraft or healing cults like *Aberewa, Hwemeso, Nana Tonga,* and *Tigari*.³¹ M.J. Field pointed out that between 1910 and 1925, when Ghana's cocoa production increased nearly tenfold from 21,000 tons to 206,000 tons, several of these healing cults were established.³² This type of cult in southern Ghana did not originate with the cocoa industry, however, and the first historically documented healing cult in southern Ghana was *Domankama* in 1879.³³ While healing cults did not originate with the cocoa industry, the social instability that accompanied the accumulation of individual wealth resulted in a pluralist ritual economy characterized by innovation and an attraction to exotic supernatural powers.³⁴ T. C. McCaskie writes that healing cults responded to social instability resulting from economic individualism and the accumulation of individual wealth, construed by many in Ghana as a threat to society.³⁵

Exotic supernatural powers were also found within new forms of Christian therapy, particularly within Faith Tabernacle and its Pentecostal branches. The rise of these enchanted churches within the Basel Mission's sphere of influence coincided with three interrelated events, which created a more immediate crisis for many within the Basel Mission community in 1918.

Crisis in the Basel Mission in 1918

While the cocoa economy and witchcraft were wreaking social havoc in Ghana, three interrelated events in 1918 led to an immediate

---

31. There is a plethora of adjectives used in the scholarly literature that describes the same phenomena, which I call healing cults. Antiwitchcraft shrines are also referred to as protective shrines, medicine eating shrines, or *suman* shrines. Alternatively, these shrines are referred to as movements or cults. For a detailed discussion of healing cults, see McCaskie, "Anti-Witchcraft Cults"; and Field, *Search for Security*, 87–104. See also Allman and Parker for a detailed history of *Nana Tonga* in southern Ghana (*Tongnaab*, 106–42).

32. Field, *Search for Security*, 29.

33. Allman and Parker, *Tongnaab*, 126–27; McCaskie, "Anti-Witchcraft Cults." The deities of the antiwitchcraft shrines probably correspond to the ɔkomfo bosom described by Christaller in the nineteenth century, which he translates as being a soothsayer's or medicine man's demon that was consulted by individuals in times of sickness or misfortune (*Dictionary*, 598).

34. Allman and Parker, *Tongnaab*, 123.

35. McCaskie, "Anti-Witchcraft Cults," 132–33.

crisis within the Basel Mission in Ghana: the Basel missionaries were expelled from Ghana, the deadly second wave of the 1918–19 influenza epidemic erupted, and cocoa production significantly declined while its price on the world market plummeted. All three crises stemmed directly or indirectly from events surrounding World War I.

Basel Missionaries Expelled

By February 2, 1918, near the end of World War I, all the German and Swiss Basel missionaries were expelled from Ghana owing to British views that German-speaking nationals were potential spies. Two days later, the Basel Mission Trading Company was liquidated; the property and stock of the company was given to the custodian of enemy property. Later, the Commonwealth Trust Limited was set up to continue this commercial enterprise.

In response to the expulsion of the German-speaking missionaries from Ghana, the British invited missionaries from the United Free Church of Scotland, who had been working in Calabar, Nigeria, to lead the Basel Mission. These Scottish Presbyterians organized the church politically on Presbyterian lines (which is ultimately how the Basel Mission became a Presbyterian church). On August 14, 1918, the first synod of the Presbyterian Church of Ghana met in Akropong, a meeting that included all the Ghanaian leadership and the Scottish missionaries. At this meeting Ghanaians were voted into leadership positions and the executive synod committee, comprised of eight Ghanaians and three Scottish missionaries, became the major decision-making body of the church.[36]

This event, which empowered male Ghanaian leadership, also created a great sense of insecurity when the Basel missionaries, who had helped manage the church and related institutions for the previous ninety years, were quickly expelled from the country. The sudden expulsion of the Basel missionaries, writes Paul Jenkins, produced a crisis of consciousness within the Basel Mission, as the organization was unprepared for such an abrupt change.[37] Two weeks after this meeting in Akropong, the influenza pandemic struck Ghana.

---

36. Smith, *Presbyterian Church*, 160.
37. Jenkins, "Afterward," 199.

Influenza Pandemic Erupted

The deadly second wave of the 1918–19 influenza pandemic emerged in Brest, France, a major Atlantic port and landing point for American troops, who in turn carried the virus all over the world.[38] On August 31, 1918, influenza outbreaks were recorded in Cape Coast, and three days later in Accra. From the coast, the influenza virus spread quickly up the rail line to Asante and then slowly into the Northern Territories by early November. By January 1919, almost every village in Ghana had been infected with influenza. Mortality in Ghana (4 percent) was higher than the average for the African continent (1.5 percent): approximately one hundred thousand Ghanaians died, mostly between October and December 1918.[39]

No institutionalized form of therapy or religion in Ghana could successfully combat the influenza pandemic. The colonial physicians in Ghana did not know what caused the pandemic, how to arrest the virus's progress, or how to treat its victims. Physicians in Ghana were not only unprepared to treat the infected, but their facilities were also already severely understaffed because of the need for physicians in Europe to treat injured soldiers during World War I. Akan healers, too, were reported to be ineffective.[40] The mainline churches did not offer any form of spiritual protection against the virus. Most churches and related institutions, such as the Basel Mission Teachers Training College, for instance, were closed by government order.[41]

Cocoa Production Halted and Prices Declined

The health crisis wrought by the "Spanish flu," as it was colloquially known, severely restricted cocoa production because the virus primarily afflicted the (scarce) labor force. Young men and women aged twenty to forty, who were primarily involved in cocoa growing and processing, had the highest mortality rates. Also, with the February liquidation of the Basel Mission Trading Company, which was responsible in part for purchasing and transporting harvested cocoa, the transportation of harvested cocoa to the coast for sale was severely

---

38. Patterson and Pyle, "Geography and Mortality," 5–6.
39. Patterson, "Demographic Impact," 410, 419.
40. Ibid.
41. Ibid., 492; Smith, *Presbyterian Church*, 159.

restricted. At the height of the pandemic, ships meant to carry cocoa neglected to stop at the major port in Accra.[42] Subsequently, the virus significantly decreased the amount of cocoa exported in 1918. Only 66,000 tons of cocoa was exported in 1918, as opposed to 91,000 in 1917 and 176,000 in 1919. The huge increase in cocoa exported in 1919 was to an extent a result of nonexported cocoa being shipped the following year, as the height of the pandemic corresponded to the beginning of the cocoa marketing season.[43]

The quantity not only diminished in 1918, but Ghanaian cocoa was also severely devalued that year on the world market. Cocoa prices on the world market continued to rise through 1916. But in 1917, the United Kingdom cut cocoa imports by half, and in late 1918 the United States prohibited the importation of Ghanaian cocoa because of the inferiority of some shipments.[44] These events caused the decline in Ghanaian cocoa prices on the world market. In 1918, a British official report recorded that Ghanaian cocoa was receiving the worst prices in the world.[45] The average value per ton for 1918 was twenty-seven pounds sterling, while the value in 1917 was thirty-five pounds and the value in 1919 was forty-seven pounds. The financial crisis of Ghanaian cocoa farmers was acute in 1918: both the quantity of cocoa produced and its market value dropped drastically from the previous year.

Faith Tabernacle Congregation in Ghana, 1918–26

These three events—the expulsion of the Basel missionaries, the massive loss of life from the influenza pandemic, and the sharp decline of cocoa production and prices on the world market—all occurred in 1918.[46] These accumulative crises were met with a religious response among Ghanaian Presbyterians. Many young Presbyterian cocoa migrants, particularly in the Eastern and Asante Regions, joined a branch of Faith Tabernacle Congregation, a divine healing church.[47]

---

42. Patterson, "Influenza Epidemic," 488.
43. Gareth Austin, personal communication to author, August 30, 2009.
44. Hill, *Gold Coast Cocoa Farmer*, 108.
45. Ibid., 110.
46. Some of the most popular healing shrines in Asante, such as *Tigare* and *Kundi*, were established around the time of the influenza pandemic in 1918 (Field, *Search for Security*, 90).
47. Mohr, "Cocoa, Chaos, and Christian Healing."

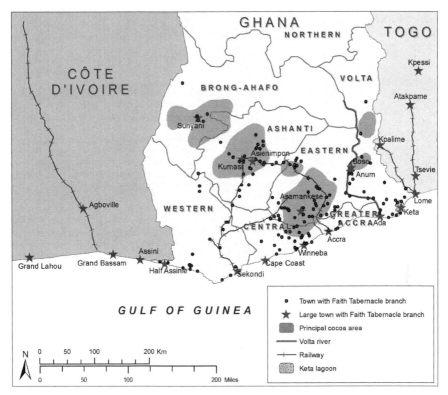

Map 2.1. Location of Faith Tabernacle branches in colonial Ghana, Côte d'Ivoire, and Togo between 1918 and 1926. This map has been drawn with postcolonial political boundaries, showing the contemporary regions of Ghana to demonstrate the quantity of cocoa production and Faith Tabernacle activity in the contemporary Eastern and Ashanti Regions, where the Presbyterian Church was most influential. Map by Zeljko Rezek.

While Ghana's cocoa fields were particularly fertile grounds for this enchanted church, Faith Tabernacle's network spread from the flagship branch in Winneba to Côte d'Ivoire in the east and Togo in the West (see map 2.1).

Faith Tabernacle Origins, Theology, and Missions

Faith Tabernacle was established in 1897 by Jacob Thomas Wilhide, a leader in John Alexander Dowie's Christian Catholic Church, which was the largest divine healing church in the United States at the turn of the century.[48] The first Faith Tabernacle pastor was John Wesley Ankins, who worked in conjunction with Wilhide in Philadelphia, but led a group that met separately.[49] Ankins's group, originally called Gospel Mission, was formally affiliated with the Christian Catholic Church until about 1899, when it cut ties with Dowie's church.[50] In February 1904, Gospel Mission purchased its first church building in Philadelphia at 2738–40 North Second Street, after which it changed its name from Gospel Mission to Faith Tabernacle Congregation.[51]

Faith Tabernacle's theology and practice of divine healing was very similar to that of Johann Christoph Blumhardt. The devil was the cause of disease and misfortune, while demons directly afflicted the suffering.[52] Through demonic possession and affliction, sickness and antisocial behavior were manifest.[53] Disease was enabled by personal sin, and one particularly sinful behavior was idolatry, a category that included alternative religious healing traditions such as spiritualism, Russelism (Jehovah's Witnesses), and Christian Science, as well as nonspiritual medical traditions such as allopathy, homeopathy, osteopathy, and chiropractic medicine.[54] Repentance of one's sins was a necessary step in receiving healing.[55] Jesus had the capacity to eradicate illness and misfortune, as well as sin, for the true believer.[56] Through prayer, laying on of hands, and anointing with oil, demons were cast out and healing achieved.[57]

---

48. For a complete account of Faith Tabernacle's early domestic history, see Mohr, "Out of Zion."

49. John Wesley Ankins, "Obeying God in Baptism," *Leaves of Healing* 5, no. 19 (March 4, 1899): 364.

50. Zehring, "Our Church History."

51. Ambrose Clark purchased this property as a trustee for Gospel Mission on February 15, 1904. Philadelphia Office City of Land Transfer and Deeds, *2738–2740 N. Second Street*, Department of Records, City Archives, City of Philadelphia. The first known issue listing Faith Tabernacle as the name of the church is *Sword of the Spirit* 4, nos.12–1 (April–May 1904).

52. *Sword of the Spirit* 1, no. 3 (July 1901): 6.

53. Spirit possession was a recognized affliction within the church during this period. For instance, one early issue of *Sword of the Spirit* featured the deliverance of a girl who was demonically possessed. *Sword of the Spirit* 1, nos. 9–10 (January–February 1902): 6.

54. *Sword of the Spirit* 22, no. 4 (n.d.): 1–2; *Sword of the Spirit* 1, nos. 9–10 (n.d.): 5; *Sword of the Spirit* 24, no. 6 (n.d.): 1–2; *Sword of the Spirit* 22, no. 10 (n.d.): 7.

55. *Sword of the Spirit* 18, no. 10 (n.d.): 5–6.

56. *Sword of the Spirit* 20, no. 2 (n.d.): 1–2.

57. Testimony of Margaret Foy Nagler (Ankins, *Words of Healing*, 123–26).

Faith Tabernacle theology was expounded in its literature, particularly the church's monthly periodical *Sword of the Spirit*, which was first published in May 1901. The production and distribution of Faith Tabernacle literature was accompanied by a distinct missionary policy. By 1903, *Sword of the Spirit* was encouraging its readers to become "home missionaries," suggesting they should send in the names and addresses of friends interested in Christianity to be put on the church mailing list.[58] By 1908, Faith Tabernacle established a Missionary Department.[59] The missionary objective of the church was not to send missionaries into other parts of the world as most churches had done, but rather to have followers around the world receive monthly supplies of Faith Tabernacle literature to distribute locally.[60] By late 1923, Faith Tabernacle was distributing literature and corresponding with followers in the West Indies, Hawaii, Alaska, the Philippines, many South American countries, England, Ireland, France, Germany, Bulgaria, Russia, Norway, India, China, Sri Lanka, Malaysia, Borneo, Fiji, Micronesia (Yap), and Australia, as well as Ghana and Nigeria.[61]

Because other British and American evangelical literature began to be imported to Ghana in bulk only after 1914, Faith Tabernacle was one of the very first—if not the first—evangelical church to disseminate its literature in Ghana.[62] The first pieces of Faith Tabernacle literature to reach Ghana were sent in the first decade of the twentieth century to European or American mainline missionaries, who in turn distributed them to Ghanaians in their churches and schools.[63] In August 1906, the first testimony from Ghana was published that referred to nonspecific "blessings" as a result of reading *Sword of the Spirit*, after which the author claimed to have distributed the periodical among his friends.[64]

Many Ghanaian Christians—particularly within the Basel Mission/Presbyterian Church—came into contact with Faith Tabernacle literature through their church or church school, and found the message of divine healing attractive.[65] Some wrote to the central branch

---

58. *Sword of the Spirit* 3, nos. 8–9 (December 1903–January 1904): 3.
59. *Sword of the Spirit* 8, nos. 7–8 (November–December 1908): 2.
60. *Sword of the Spirit* 9, no. 5 (September 1909): 7.
61. *Sword of the Spirit* 21, no. 3 (n.d.): 3.
62. Newell, *Literary Culture*, 87.
63. "For the Glory of God," *Sword of the Spirit* 21, nos. 2–3 (n.d.).
64. "Publishing House Notes," *Sword of the Spirit* 6, no. 6 (October 1906): 4.
65. Divine healing literature was circulating among mainline Christians in Ghana before Faith Tabernacle began to rapidly expand after 1918. A small following of

in Philadelphia for more copies of *Sword of the Spirit*, sermons, and tracts, and began evangelizing locally as the leadership in Philadelphia had suggested in creating "home missionaries."[66] Before formal branches of Faith Tabernacle were established in 1918, the divine healing church functioned as a parachurch organization: an extradenominational Christian institution that distributed its literature and had followers within various denominations.

But why would Basel missionaries distribute another church's literature? At the turn of the twentieth century, the divine healing movement was very interdenominational. The Basel Mission Book Depot at times sold religious material written by divine healing proponents such as Reuben A. Torrey.[67] Some Basel missionaries like Elias Schrenk were divine healing proponents. The divine healing movement had begun among Württemberg Pietists, who made up the majority of Basel missionaries in Ghana. As Faith Tabernacle had a significant amount of German-speaking members (and Philadelphia was the center of Württemberg immigration to the United States),[68] it is probable that Faith Tabernacle members would have had family and friends involved in the Basel Mission, which was primarily made up of Württembergers.

Influenza Pandemic and the Beginning of Faith Tabernacle in Ghana

In Ghana, the first Faith Tabernacle congregation was established in Winneba sometime in the first half of 1918, before the influenza pandemic erupted in September. Two brothers, Joel Sackey Sam and Josephus Kobina Sam, were walking one morning toward the

---

John Alexander Dowie's Christian Catholic Church, led by the Methodist Elder D. Kwesi Bhoma, was based in Axim between 1901 and 1905. Bhoma was receiving issues of Dowie's periodical *Leaves of Healing*, as well as publishing articles in this journal: Bhoma, "African and Son Healed of Smallpox," *Leaves of Healing* 9, no. 25 (October 12, 1901): 809.

66. "For the Glory of God," 3.

67. Newell, *Literary Culture*, 11; Gloege, *Moody Bible Institute*, 41–71.

68. Faith Tabernacle's German-speaking members from Württemberg were so numerous that by 1901 Faith Tabernacle began writing tracts in German and employed a German-speaking evangelist, H. Rudolph Otto. *Sword of the Spirit* 1, no. 3 (July 1901): 4. There is evidence that extensive communication existed between Basel missionaries and American relatives; one source claims that religious issues in Ghana were discussed and debated across the Atlantic Ocean between former Basel missionaries in Germany and their relatives in the United States (Debrunner, *History of Christianity*, 325).

Muni Lagoon and found a copy of a Faith Tabernacle sermon on the ground partly covered in mud. After becoming interested in the sermon's message, the two brothers began corresponding with Pastor Ambrose Clark in Philadelphia and established a branch in Winneba.[69] Joel Sackey Sam, assisted by his brother, became the leader of this flagship Faith Tabernacle branch in Ghana.[70] The two Sam brothers, who were formerly elders in Winneba's Wesleyan Mission, started their congregation initially with some of their family and a few members of their former church.[71] Not long after this Faith Tabernacle congregation was formed, the influenza pandemic erupted.

The 1918–19 influenza pandemic catalyzed the rapid spread of Faith Tabernacle in Ghana, partly because this virus resulted in extreme suffering in both Philadelphia and Ghana. In Philadelphia, the influenza virus produced the highest mortality rate of any city in Europe or North America; nearly 1 percent of the population (almost sixteen thousand people) died between September and late December 1918.[72] Faith Tabernacle recorded the good health of its congregation in the midst of this pandemic in *Sword of the Spirit*. The headline of the October 1918 issue read, "God's Witnesses To Divine Healing: Healed of Spanish Influenza and Pleuro-Pneumonia." Inside were testimonies given by two women, as well as a list of over fifty persons miraculously healed of influenza and accompanying pneumonia through their faith in God alone. Other healing testimonies from the 1918–19 influenza pandemic were recorded in the periodical the following year and continued until the late 1920s.[73] These healing testimonies were circulating in Winneba, as well as the various Christian circles throughout Ghana where Faith Tabernacle literature was read.

---

69. Okai, *General History*, 1.

70. Faith Tabernacle Congregation, Correspondence Notebooks (hereafter FTC-CN), Joel Sackey Sam to Ambrose Clark, August 1, 1924.

71. Okai, *General History*, 1. Wyllie claimed that the initial congregation was comprised of seventeen members (*Spirit Seekers*, 23).

72. Barry, *Great Influenza*, 197–99.

73. "Healed of Spanish Influenza and Pluera-Pneumonia," *Sword of the Spirit* 17, no. 8 (October 1918): 1–2; "Testimony of Margaret Henry," *Sword of the Spirit* 18, no. 8 (n.d.): 2–3; "Testimony of Mrs. Sarah B. Duffy," *Sword of the Spirit* 18, no. 10 (n.d.): 2; "Testimony of Mr. and Mrs. Edward Proudfoot," *Sword of the Spirit* 22, no. 10 (n.d.): 3; "Testimony of Miss Natalie D. Wurts," *Sword of the Spirit* 24, no. 3 (n.d.): 2.

The influenza pandemic caused the worst short-term demographic disaster in the twentieth century; approximately one hundred thousand Ghanaians died, mostly between October and December 1918.[74] No institutionalized form of therapy or religion in Ghana could successfully combat the influenza pandemic. Faith Tabernacle, however, responded proactively. At the peak of the pandemic in October and November, many people were brought from the nearby Winneba hospital to Faith Tabernacle's divine healing home, called Faith Home, which was a place for the sick to rest, be prayed for, and heal.[75] As with Philadelphia survivors recorded in *Sword of the Spirit*, many Ghanaian survivors testified to their miraculous healings from the virus without the use of any medication. The ability of the church to heal, in contrast to the hospital's inability to do so, looms large in the memories of the descendants of the Sam brothers in Winneba.[76]

After the influenza pandemic struck Winneba, small congregations began to emerge on the outskirts of Winneba and later in various coastal towns.[77] Fante migrants from Winneba established branches as far away as Côte d'Ivoire and northern Nigeria.[78] But a significant amount of the evangelizing in Ghana was done by a former Basel Mission elder from Asienimpon (outside Kumasi) named Kwadjo Nti, who spread Faith Tabernacle's doctrines and healing practices around all major ethnic regions of southern Ghana and into Togo from the flagship congregation in Winneba.[79] By May 1920, Ambrose Clark reported a great revival in West Africa (Ghana and Nigeria),

---

74. Patterson, "Influenza Epidemic," 502.

75. The original Faith Home was established in October 1907 in Philadelphia in a building adjacent to the Faith Tabernacle sanctuary ("Faith Tabernacle Home," *Sword of the Spirit* 7. no. 6 [October 1907]: 4). Other divine healing homes were established in Dunkwa by Joseph Addo and in Kwadwowusu by E. B. Apemah (FTC-CN, Joseph Addo to Ambrose Clark, July 15, 1924; E. B. Apemah to Ambrose Clark, January 25, 1925).

76. Interview with Kenneth Gyasi Sam (eldest son of Josephus Kobina Sam), Winneba, February 16, 2007.

77. Ibid.

78. FTC-CN, I. G. Hayford to Ambrose Clark, October 15, 1924; S. A. Mensah to Ambrose Clark, September 9, 1924, and December 22, 1924; Dan K. Turkson to Ambrose Clark, December 3, 1924.

79. Faith Tabernacle records in Philadelphia refer to Kwadjo Nti as Timothy Anti (Okai, *General History*, 2–5). At this time, in 1918, southern Togo was under British authority, which could have facilitated the circulation of Christian literature and evangelists like Nti into Togo.

which included fifteen or twenty branches, some with several hundred members. The majority of these branches were in Ghana, not Nigeria; Ghana's Faith Tabernacle membership was nearly four times larger than Nigeria's in the first half of the 1920s.[80] Ambrose Clark wrote that the Africans with whom he corresponded were receiving the "Full Gospel" of salvation and healing from diseases by the power of God alone. Many Ghanaian followers had their testimonies published in *Sword of the Spirit*. For instance, a Ghanaian in Kano, Nigeria, wrote in August 1920:

> I don't know why our missionary here did not teach us the full gospel.... They never did wondrous work here as our Lord give the power to the disciples. Should you bring your Divine Healing church out in Africa here and preach the full gospel to the people and healing the sick, many people shall believe in your Gospel than those churches who were teaching us here, because they never done miracle; they never heal the sick.[81]

Faith Tabernacle's healing practices and theology strongly resonated with mainline Christians who were committed to their faith but had no ritual means of addressing their illnesses or misfortunes. By the end of 1923, Ambrose Clark claimed to have distributed thousands of pieces of literature and was receiving up to two hundred letters a day from Ghana and Nigeria.[82]

By early 1926, there were at least 177 Faith Tabernacle branches in Ghana that extended east into Togo and west into Côte d'Ivoire (see map 2.1); 115 leaders from these branches corresponded with the headquarters in Philadelphia. Leaders reported congregations in 1925 as large as 600 (Half Assini) and 369 (Mangoase).[83] With a

---

80. There were 177 branches in greater Ghana compared with only 46 Nigerian branches. And Ghanaian migrants from Winneba established many of the northern Nigeria branches, such as Jos, Kaduna, and Minna. Ghana's estimated membership was 4,425 compared with approximately 920 in Nigeria by 1926.

81. "God's Blessings Continued," *Sword of the Spirit* 19, no. 2 (n.d.): 4, 7.

82. "For the Glory of God," *Sword of the Spirit* 21, no. 3 (n.d.): 2–3.

83. In the letters mailed to Ambrose Clark, Ghanaian leaders frequently gave statistics with regard to the number of membership forms completed, the membership number of a particular branch, or the attendance in a particular congregation on a particular day. Some of the larger numbers include: over 600 members at Half Assini (FTC-CN, Joel Sackey Sam to Ambrose Clark, September 28, 1925) and 369 members at Mangoase (FTC-CN, Joel Sackey Sam to Ambrose Clark, February 24, 1925).

conservative estimate of 25 members per branch, there were approximately 4,425 Faith Tabernacle members in greater Ghana by 1926. In that same year, the Presbyterian Church—established nearly one hundred years prior (in 1828)—claimed roughly 22,000 members in 260 congregations.[84] By comparison, Faith Tabernacle garnered 20 percent of the Presbyterian Church's membership (as well as several of its members) worshipping in over 68 percent of its congregations.

Faith Tabernacle in the Cocoa Fields

Faith Tabernacle was predominately a religion of migrant Christian wage-earners, and its branches grew exponentially in the cocoa growing regions of Ghana, particularly among Presbyterians. Faith Tabernacle branches in Ghana were found in the largest cocoa-producing regions, such as Akuapem and Akyem (thirty-four branches), Asante (twenty-three branches), Kwahu (seven branches), greater Sunyani (five branches), and in and around Anum and Boso (four branches). More than 40 percent of the Faith Tabernacle branches were founded in cocoa-growing regions in the interior. Many of these Faith Tabernacle leaders were schooled in the Basel Mission and could have either worked as clerks or farmers within the cocoa industry.[85] Other leaders, like James Nkansah of Anyinam and W. A. Johnson of Mumford, were solely cocoa farmers.[86] J. C. Isaiah, a Faith Tabernacle leader in Agboville, Côte d'Ivoire, owned his own cocoa-purchasing firm called

---

Faith Tabernacle membership was not always as exclusive as Ambrose Clark wished, as many members did not fully separate from their prior churches. In 1924, for instance, A. E. Fiagbedzi was leading a Faith Tabernacle congregation in Keta, while still working as a teacher and temporary pastor in the African Methodist Episcopal Zion Mission School (FTC-CN, A. E. Fiagbedzi to Ambrose Clark, October 8, 1924).

84. Smith, *Presbyterian Church*, 292.

85. While his exact profession in Asamankese (besides pastor) is unknown, Peter Anim could have been a clerk or a farmer. Farms, particularly if owned by a group or company, needed clerks to work as secretaries or accountants (Hill, *Migrant Cocoa-Farmers*, 41).

86. FTC-CN, James K. Nkansah to George W. Foster, September 25, 1935; W. A. Johnson to Ambrose Clark, January 29, 1925. Another pastor, Isaac Ampomal, wrote that his prayers for privately owned farmland in Akyem Manso were answered (FTC-CN, Isaac Ampomal to Ambrose Clark, March, 28 1925). One reference in Faith Tabernacle's Minutes Book from 1942 describes a Faith Tabernacle pastor who was free to work more for the church while his cocoa trees matured. Faith Tabernacle Congregation Ghana, *Minutes Book* (1936–43), 169.

Isaiah Brothers Ltd.[87] Mirroring the cocoa economy with respect to gender, all leaders of Faith Tabernacle were men, as women were not allowed positions of leadership within the church.

The most significant Faith Tabernacle leader in the cocoa fields was Peter Anim, who was arguably the second most powerful Faith Tabernacle pastor in Ghana during the 1920s after Joel Sackey Sam.[88] Peter Anim established a thriving branch of Faith Tabernacle in the cocoa migrant town of Asamankese, in the heart of Akyem Abuakwa, and garnered nine other branches under his authority in Amanese, Anum, Apeso-Kubi, Asuboi, Boso, Finte, Kpesse, Kwadwowusu, and Pese.[89]

Peter Anim was originally from the adjoining towns of Anum and Boso on the eastern bank of the Volta River. He attended the Basel Mission school in Boso and continued his education at the Basel Mission secondary school in Anum where he graduated in 1908 (see fig. 2.2.).[90] After graduating, Anim was employed with the Basel Mission Factory in Pakro as a weighing clerk.[91] In 1917, while living in Boso, Anim obtained a copy of *Sword of the Spirit*.[92] Impressed with the church's divine healing message, Anim became a subscriber. In 1920, Anim's first wife died, after which he began to follow Faith Tabernacle teachings. In 1921, Anim contracted guinea worm and wrote to Ambrose Clark, who encouraged Anim to have faith and trust God for his healing. Anim was later healed of this parasitic infection along with chronic stomach pain. After his healing, Anim withdrew his membership from the Basel Mission at Boso and settled in Asamankese, following a massive labor migration of Anum and Boso residents who relocated to work in the cocoa industry beginning as early as 1907.[93]

Anim's Faith Tabernacle branch became wildly successful in the first half of the 1920s. By 1921, Peter Anim began evangelizing in Asamankese and was soon joined by several converts. A Basel minister's report from Asamankese two years later lamented that their

---

87. FTC-CN, J. C. Isaiah to Ambrose Clark, April 4, 1925.
88. Anim frequently helped to resolve conflicts in the Faith Tabernacle community with Joel Sackey Sam, particularly in the Nsawam branch. FTC-CN, J. M. Ammah to Ambrose Clark, December 6, 1924; Peter Anim to Ambrose Clark, February 10, 1925.
89. FTC-CN, Peter Anim to Ambrose Clark, February 10, 1925.
90. Larbi, *Pentecostalism*, 99.
91. Ibid.
92. Anim, *History*, 1.
93. Debrunner, *History of Christianity*, 254; Hill, "Migrant Cocoa Farmers," 250.

Figure 2.2. "Secondary School in Anum, 1907." This photograph of the Basel Mission's Secondary School in Anun was taken a year before Peter Anim's graduation. Anim is probably pictured. Ref. no. D-30.12.011. Photograph by Wilhelm Erhardt. Reproduced with permission from the Basel Mission Archives / Basel Mission Holdings.

members were joining Faith Tabernacle in droves.[94] In October 1923, Anim received a certificate of pastoral authority from Ambrose Clark giving him the right to baptize and confirm appointments. Between December 28, 1923, and January 2, 1924, Anim attended the annual meeting of Faith Tabernacle pastors in Winneba in order to become more integrated into the larger network of churches.[95] In a letter to Ambrose Clark written in July 1924, Anim reported nearly 100 new members and another 122 recent baptisms in his Asamankese church, which already had a substantial following.[96]

By the mid-1920s, Faith Tabernacle had become a serious problem for the Presbyterian Church, which dealt with this enchanted competitor in various ways. In 1923, the Presbyterian Synod Committee concluded that the church must deal gently with the Presbyterian followers of Faith Tabernacle, but if the followers refused to support

---

94. Smith, *Presbyterian Church*, 256.
95. Anim, *History*, 3.
96. FTC-CN, Peter Anim to Ambrose Clark, July 19, 1924. Baptism did not automatically confirm membership.

the Presbyterian Church, they would be excommunicated.[97] And leaders of the Presbyterian Church did excommunicate Faith Tabernacle followers.[98] One Presbyterian pastor intimidated Faith Tabernacle followers of his congregation with the threat that burial in the chapel with sacred mission rites would be denied them.[99] Another Presbyterian pastor in Asuboi took the Faith Tabernacle contingency of his congregation to court because they would not remove themselves from their communally held land.[100] And yet another Presbyterian pastor tried to petition the colonial authorities to prevent Faith Tabernacle's Koforidua branch from holding meetings.[101] Finally, a Presbyterian pastor in Nyakrom prevented the distribution of Faith Tabernacle literature within his congregation.[102]

Faith Tabernacle to the Apostolic Church, 1926–38

Faith Tabernacle was growing rapidly in Ghana until a major division occurred in Philadelphia in late 1925. In October of that year, Ambrose Clark left Faith Tabernacle to form the First-Century Gospel Church, also in Philadelphia.[103] While this fact is uncontested, the account of events leading up to Clark's departure has two versions. The Faith Tabernacle version claims that Ambrose Clark committed adultery on several occasions and was replaced by George W. Foster as presiding elder.[104] Alternatively, the First-Century Gospel version argues that Ambrose Clark was on vacation when George W. Foster secretly called a meeting of Faith Tabernacle pastors to take control

---

97. Presbyterian Church, Synod Committee meeting minutes, no. 48 (July 1923), in Larbi, *Pentecostalism*, 153.
98. FTC-CN, E. B. Apemah to Ambrose Clark, January 25, 1925.
99. FTC-CN, Robert F. Poe to Ambrose Clark, December 16, 1925.
100. FTC-CN, Peter A. Amoako to Ambrose Clark, April 16, 1925; Samuel Benoa to Ambrose Clark, May 8, 1925; Isaac Ampomal to Ambrose Clark, March 28, 1925.
101. FTC-CN, James Kingston Tsagli to Ambrose Clark, September 20, 1924.
102. FTC-CN, Samuel Benoa to Ambrose Clark, May 18, 1925.
103. By October 22, 1925, Ambrose Clark was no longer Presiding Elder of Faith Tabernacle Congregation. Faith Tabernacle Congregation, Board of Elders minutes (October 1925). A notice in *Sword of the Spirit* stated that since October 22 all letters sent to Ambrose Clark of Faith Tabernacle were received by Clark at his home, since he was no longer pastor of Faith Tabernacle. "Announcement," *Sword of the Spirit* 23, no. 4 (October 1925): 4.
104. FTC-CN, Presiding Elder George W. Foster to Pastors of the Faith Tabernacle, October 29, 1925.

of the church.[105] Both versions concur that Pastor Clark was accused of adultery by the Faith Tabernacle board of elders and his leadership was revoked. As a result, Ambrose Clark, his family, and a large portion of the Faith Tabernacle congregation left to form the First-Century Gospel Church.

The two churches fought over both local and West African branches. In the local courts, Ambrose Clark and the Faith Tabernacle leadership fought over the incoming mail addressed to "Pastor Clark of Faith Tabernacle." The Philadelphia courts ruled in favor of Clark.[106] Faith Tabernacle mailed a circular to domestic and foreign branches of the church explaining its version of the split to pastors in late October and to congregants in early December.[107] By the time these letters arrived in Ghana, many of the Faith Tabernacle congregations had followed Ambrose Clark and changed their name to the First-Century Gospel Church.

In late 1926, Ambrose Clark wrote in his new monthly periodical *First-Century Gospel* about his church's large following in Ghana.[108] This following included Joel Sackey Sam in Winneba, who in joining First-Century Gospel destroyed the major Faith Tabernacle network, as Winneba was the central node. Most of the correspondence between Faith Tabernacle in Philadelphia and branches in Ghana dwindled rapidly after 1926. Faith Tabernacle went from approximately 4,400 members in 1926 to 58 total members in 1963.[109]

After Ambrose Clark's untimely departure in late 1925, many Faith Tabernacle branches, leaders, supporters, and affiliates—drawn to the doctrine of baptism in the Holy Spirit by the 1920s—formed the first Pentecostal churches in Ghana in the 1930s.[110] Prior to Clark's depar-

---

105. Pastor A. Clark to Foreign Correspondence, n.d., FTC-CN. Also, Pastor James Clark (Ambrose Clark's grandson), conversation with author, New Jersey, March 4, 2007.

106. Pastor James Clark, conversation with author, New Jersey, March 4, 2007.

107. FTC-CN, Faith Tabernacle Congregation to Pastors of Faith Tabernacle, October 29, 1925; Presiding Elder George W. Foster to correspondence of Faith Tabernacle, December 4, 1925.

108. "Testimony of Pastor A. Clark and Family," *First-Century Gospel* 1, no. 9(October 1926): 6.

109. FTC-CN, Samuel A. Baidoo to Charles A. Reinert, May 15, 1963.

110. Anim, *History*, 3–4. There is some evidence that Pentecostal influences existed in Ghana before the 1920s and outside of Faith Tabernacle circles. For instance, in 1910 the British Pentecostal periodical *Confidence* was distributed in Ghana. *Confidence* 3, no. 12 (December 1910): 284. Also, a Liberian Pentecostal from

ture, in 1924, a Faith Tabernacle member near Keta asked Ambrose Clark about the "tongues movement."[111] In 1925, another Faith Tabernacle member reported Pentecostal groups in Agona and Fante.[112] By 1928, Peter Anim was reading the periodical of the Apostolic Faith and becoming interested in this American-based Pentecostal church.[113] By June 1930, Anim and his congregation—joined by former Faith Tabernacle congregations from Akenkansu, Aperade, Asienempon, Ayirebi, and Oda—had fully embraced the Pentecostal movement.[114] One year later in 1931, missionaries from the British-based Apostolic Church visited Peter Anim. As a result, in 1935, Anim and much of his substantial following joined this Pentecostal church.[115] From Asamankese, Pentecostalism spread through their network of current and former Faith Tabernacle and First-Century Gospel associates to Anlo, Asante, Fante (including Winneba), Krobo, and into Togo.[116]

The union between Anim and the Apostolic Church was short-lived, however. By the middle of 1938, the Apostolic Church in Ghana split over a theological issue. One side, led by Peter Anim in Asamankese, abstained from all forms of medicine, while the other, led by the British missionary James McKeown in Winneba, incorporated biomedicine.[117] The two major nodes of the Faith Tabernacle network in the 1920s—Winneba and Asamankese—had become the two major Pentecostal nodes in the 1930s, and Pentecostalism continued to advance in Ghana among former Faith Tabernacle leaders, congregants, and affiliates.[118] A focus on suffering and a ritual repertoire

---

Cape Palmas named Jasper K. Toe reported evangelizing in Ghana during 1918 in the midst of the influenza pandemic. "How Missions Pay," *The Latter Rain Evangel* 11, no. 7 (April 1919): 13.

111. FTC-CN, Albert V. C. Nanevie to Ambrose Clark, November 17, 1924. Literature condemning Pentecostalism or the "tongues movement" had been circulating within Faith Tabernacle circles in Ghana since at least the mid-1920s. FTC-CN, A. M. Boateng to Ambrose Clark, February 7, 1925.

112. FTC-CN, Jacob R. Mensah to Ambrose Clark, May 8, 1925.

113. Anim, *History*, 3.

114. FTC-CN, E. Edward Brown to Edwin Winterborne, January 26, 1931. On August 4, 1930, Pastor Brown from Accra revoked Peter Anim's Faith Tabernacle credentials and mailed them back to Philadelphia.

115. Anim, *History*, 4, 6–7.

116. Ibid., 6.

117. Bredwa-Mensah, "Church of Pentecost," 11.

118. This supports Maxwell's argument that the key to Pentecostalism's rapid global advance was the evangelical networks previously established (Maxwell, *African*

of biblically based healing methods was maintained in the transition from Faith Tabernacle to Pentecostalism in Ghana. Another feature maintained in this religious transition was patriarchy, in which men continued to occupy most if not all positions of leadership within early Pentecostalism in Ghana.

Because the McKeown-led Apostolic Church grew much more rapidly than the Anim-led Christ Apostolic Church, particularly in the Presbyterian stronghold of the Eastern Region, I will focus on the Apostolic Church in the following section.

The Apostolic Church under McKeown, 1938–52

An important factor responsible for the growth of the Apostolic Church in the 1940s was the presence in many parts of the country of former Faith Tabernacle members, followers, and entire congregations, many of whom transferred their allegiances to the McKeown-led Apostolic Church.[119] The Apostolic Church grew faster in the Eastern Region than in any other in the 1940s, and by the end of 1947 there were assemblies in Abetifi, Abirem, Akropong, Akyem Tafo, Aseseeso, Jumapo, New Amanokrom, Nkawkaw, Otumi, and Suhum. In 1949, an assembly was started in Koforidua, the capital of the Eastern Region. Many of the new converts and early leaders of the Apostolic Church in the Eastern Region were former Presbyterians or former Faith Tabernacle members.[120] Many were also young cocoa migrants.

In the fifteen-year period between 1938 and 1952, the Apostolic Church grew rapidly under McKeown, in large part because of the various healing campaigns involving leaders or members of the Apostolic Church.[121] For example, in a village in Togo near the Volta Region border, a chief was healed of blindness through the prayers of an Apostolic church member. A woman from Oda was pronounced dead at the Oda hospital but, through the prayers of her sister, was brought back to life after five days. Pastor Okanta in the Eastern Region successfully prayed for and healed a crippled boy, an insane woman, many barren women, and a woman with a strange skin condi-

---

Gifts of the Spirit, 35).
   119. Larbi, *Pentecostalism*, 181.
   120. Ibid., 182.
   121. Ibid., 186–89.

tion. Both laity and clergy of the Apostolic Church successfully prayed on behalf of many people stricken by various illnesses.

By 1952, the Apostolic Church had 512 congregations throughout Ghana, 53 ordained African pastors, and approximately 10,000 members.[122] By comparison, in the same year, the Presbyterian Church claimed approximately 593 congregations, 81 ordained pastors, and 40,000 adult members.[123] The Apostolic Church in 1952 claimed 86 percent of the Presbyterian Church's congregations, 65 percent of their pastors, and 25 percent of their membership.

The Latter Rain Revival and the Birth of the Church of Pentecost, 1953–66

In March 1951, Pastor Fred C. Poole, the superintendent of the Apostolic Church USA based in Philadelphia (the American branch of the UK church), sent a letter to James McKeown as well as to the leaders of the Apostolic Church in Eastern Nigeria and Western Nigeria.[124] Poole wrote that a great revival called the Latter Rain was sweeping North America and that God had told him to spread the revival to Africa. This Pentecostal revival was marked by healings and other miraculous phenomena, in contrast to the classical Pentecostal churches in North America, whose healing practices had diminished since the 1930s and 1940s.[125] Poole suggested that McKeown send him an invitation to allow a Latter Rain evangelical team to visit Ghana.

The Apostolic Church in the United Kingdom had formally rejected the Latter Rain revival and did not immediately allow the team to visit Ghana. An elder in the Philadelphia Apostolic Church in the 1950s argued that the UK church particularly took issue with the Latter Rain movement's leadership, which was headed by a woman, Pastor Myrtle D. Beall of the Bethesda Missionary Temple (an Assembly of God church) in Detroit.[126] The Ghanaian Apostolic Church was not swayed by the apprehension of the United Kingdom and peti-

---

122. Bredwa-Mensah, "Church of Pentecost," 18.
123. Smith, *Presbyterian Church*, 292.
124. "Facts of Troubles in Ghana Apostolic Church," Church of Pentecost Report, 1962, 1.
125. Riss, "Latter Rain Movement," 830.
126. Author interview with Elder John Hoene, former member of Fred C. Poole's Apostolic Church congregation, Philadelphia, August 1, 2010.

tioned McKeown to let the revival team come. The Latter Rain team was led by Thomas Wyatt, whose periodical *Wings of Healing*, which featured healing miracles during previous Latter Rain revivals, was popular in Ghana.[127]

In early 1953, the Latter Rain evangelists came to Ghana. The Latter Rain team, consisting of Thomas Wyatt, Fred C. Poole, and Adam McKeown (James's brother), arrived in Accra on January 23 and stayed until February 13, 1953, holding services in Accra, Cape Coast, and Kumasi.[128] One Pentecostal missionary wrote, "thousands were saved, healed and many received the Holy Ghost."[129]

Upset that the Latter Rain team promoted interdenominational Pentecostal worship, the Apostolic Church Council in the United Kingdom concluded that the Latter Rain abused its privileges, and the movement was subsequently condemned.[130] James McKeown, who supported the movement and disagreed with this verdict, was dismissed from the Apostolic Church. On hearing of McKeown's dismissal from the Apostolic Church, the Council of the Apostolic Church of Ghana agreed to break ties with the Apostolic Church of the United Kingdom. The group led by acting superintendent J. A. C. Anaman passed a resolution in which they recognized McKeown as their leader, severed relationships with the British Apostolic Church, and refused to accept any other missionaries from the United Kingdom.[131] The resolution was passed in late May 1953 and the great majority of the Apostolic Church congregations in Ghana broke their alliances with the British church.

Soon thereafter, the newly independent church adopted the name Ghana Apostolic Church. McKeown returned in October 1953 with some financial assistance provided by the Latter Rain movement, possibly from Wyatt or Poole. The two factions—the Apostolic Church of Ghana and the Ghana Apostolic Church (led by McKeown)—fought over bank accounts and church property.[132] Confusion over the two similar names persisted until President Kwame Nkrumah intervened in 1962. In response, the Ghana Apostolic Church changed its name to the Church of Pentecost on August 1, 1962.

---

127. Bredwa-Mensah, "Church of Pentecost," 19–20.
128. Onyinah, *Akan Witchcraft*, 177.
129. "Gold Coast and Nigeria," *Pentecost* 28 (June 1954): 11.
130. Larbi, *Pentecostalism*, 212.
131. Bredwa-Mensah, "Church of Pentecost," 23.
132. Ibid., 25–26.

The 1966 synod report of the Presbyterian Church of Ghana claimed that the Church of Pentecost had more adult members than the Presbyterian Church by 1966.[133] A significant reason for this was that the Church of Pentecost was enchanted; the church recognized witches and malevolent forces in the lives of its members. In particular, McKeown formally acknowledged witches and made no distinction between witches, evil spirits, and demons.[134] The Church of Pentecost also provided protection against witches, evil spirits, and demons, and a means of combating these malevolent spiritual forces. By the 1960s, the Eastern Region was no longer Presbyterian-dominated.

Challenge to Presbyterian Dominance: The Church of Pentecost in the Eastern Region

The large-scale transition from the Presbyterian Church to the Church of Pentecost has impressed the memories of current Presbyterians participating in the healing movement in Ghana that became instituted in the 1960s (see chapter 3) and in the United States (see chapters 4–6). Reverend Samuel Atiemo, pastor of the Ghanaian Presbyterian Reformed Church in Brooklyn, argues that the Church of Pentecost was started by Presbyterians who were drawn to the power of prayer and healing offered by the church.[135] Rosamund Atta-Fynn, a deacon in the Ghanaian Presbyterian Church in Philadelphia, described the older men of the Presbyterian Church in Nsaba (established in 1891), such as her grandfather, as having left the Presbyterian Church to join the Church of Pentecost because they wanted to be part of a church that had the power to meet one's needs through prayer.[136]

One of the most notable Presbyterians to take a leadership position within the Church of Pentecost was M. K. Yeboah. Yeboah was

---

133. Presbyterian Church of Ghana, Report on "Prayer Groups and Sects," Minutes of the 37th Synod, 1966, 42–43.

134. For instance, McKeown used the terms "witches" and "demons" interchangeably in a letter written in 1958 (Onyinah, "Man James McKeown," 80).

135. This interpretation was given during the introductory address at the Stony Point deliverance workshop in November 2006.

136. This information was revealed during a lecture given at the Ghanaian Presbyterian Churches in North America's annual deliverance workshop in Stony Point, New York, on November 28, 2006.

raised as a Presbyterian in Akuapem but became an overseer in the Apostolic Church by 1951.[137] He was later ordained a pastor and confirmed a prophet by the Church of Pentecost in 1954, particularly because of his gifts of healing.[138] Yeboah was chosen to be part of the first five-member executive council formed to administer the church's affairs in March 1961.[139] Later, from 1988 to 1998, Yeboah served as chairman of the Church of Pentecost. Yeboah, like Peter Anim before him, was a young Presbyterian who found a more enchanted form of Christianity: one that provided him and other young people with the power to heal and the means to combat malevolent spirits in an insecure world. As chairman, Yeboah reinstated the offices of prophet to the Church of Pentecost, giving more public recognition to healing and deliverance practices within the church.[140]

By the mid-1960s, attendance at Presbyterian services in Ghana had severely declined. While the Presbyterian community grew considerably because of incoming schoolchildren, there was very little increase in the communicant membership between 1938 and 1958.[141] In 1918, the year Faith Tabernacle became established in Ghana, Presbyterian schoolchildren comprised around one-third (34 percent) of the total Presbyterian Christian community. By 1959, schoolchildren comprised two-thirds (67 percent) of the total Presbyterian Christian community.[142] Between 1920 and 1960, the Presbyterian Church had been schooling hundreds of children, many of whom, upon reaching adulthood, joined one or more of these newer enchanted churches, such as Faith Tabernacle or the Church of Pentecost. Besides schoolchildren, the Presbyterian Church had become the church of upper-class adults, whose lives were much more secure. Many younger Presbyterians, whose lives were imminently insecure, joined enchanted churches in large numbers.

Writing critically about the Presbyterian Church of Ghana in the 1960s, Noel Smith, a former member of the Education Committee of the Presbyterian Church, lamented that the Presbyterian catechists

---

137. Council Meeting of Pastors, Overseers, and General Deacons of the Apostolic Church, Koforidua, March 19–22, 1951, 1.

138. Minutes of the Gold Coast Apostolic Church held at Kumasi, April 20–22, 1954, 1–2.

139. Onyinah, "Man James McKeown," 82, 103.

140. Onyinah, *Akan Witchcraft*, 224.

141. Smith, *Presbyterian Church*, 215.

142. Ibid., 292.

and teachers focused their efforts on moral discourses that did not satisfy the spiritual needs of members, which resulted in the large-scale conversion to Pentecostalism.[143] He further argued that the healing of the whole person—addressing a person's illness and misfortune spiritually—needed to occur in the Presbyterian Church in order for Ghana's oldest church to survive. Smith's criticisms were seriously addressed by the 1960s (see chapter 3).

Conclusion: Enchanted Competition for the Presbyterian Church of Ghana

By the turn of the twentieth century, the Basel Mission had become formally disenchanted, causing significant problems for its members. In 1918, three interrelated events caused a more immediate crisis within this Christian community: the influenza pandemic erupted, the Basel missionaries were expelled, and the production and price of cocoa drastically decreased. This crisis led many young migrant Presbyterian cocoa farmers in the Eastern Region and Asante to join Faith Tabernacle, a new church offering healing and spiritual protection from the malevolent spiritual forces that were believed to have caused the pandemic.[144] The influenza pandemic contributed to the rapid growth of Faith Tabernacle within the Presbyterian-dominated Eastern Region and within the Presbyterian community in Asante.

While the influenza pandemic initially corresponded to the rapid expanse of Faith Tabernacle within Presbyterian migrant cocoa-farming communities, it was the social upheaval wrought by the cocoa industry from the 1920s through the 1950s that fueled fears of witchcraft and stoked suspicions that demons lay behind people's illnesses and misfortunes, and these matters were addressed by Faith Tabernacle and its Pentecostal branches. The Presbyterian Church was disenchanted and offered healing only through biomedical means, which were popularly believed to be ineffective in combating witchcraft and other malevolent spiritual forces. By the 1960s, the growth of the Church of Pentecost—a Pentecostal branch of Faith Tabernacle—was significant, particularly in the Eastern Region. So many adults left the Presbyterian Church to join newer Pentecostal churches that the Presbyterian leadership decided a radical change

---

143. Smith, *Presbyterian Church*, 245–67.
144. *Sword of the Spirit* 17, no. 8 (n.d.): 1.

needed to be made. If it was to survive, the Presbyterian Church needed to become enchanted: it had to acknowledge spiritual explanations of health and suffering in the world, recognize spiritual afflictions in the lives of the church's congregants, and afford spiritually afflicted congregants healing by charismatic individuals supported by the church. It is at this historical juncture, in the early 1960s, that the next chapter begins.

3

The Enchantment of the Presbyterian Church of Ghana, 1960–2010

By the early 1960s the Presbyterian Church of Ghana was no longer the preeminent church in Ghana's Eastern Region, the area dominated by the Presbyterian Church of Ghana since the Basel Mission established its headquarters in Akropong in 1835. The primary reason was that members, particularly younger wage-earners, had been leaving the Presbyterian Church of Ghana in droves over the previous forty years. These young men and women became attracted to more enchanted forms of Christianity, churches that offered robust healing practices and combated the malevolent forces that caused these Ghanaians great suffering.

This chapter explains how, as a response to this mass exodus of members among other factors, the Presbyterian Church of Ghana became enchanted after 1960. Before I describe in detail the context in which the Presbyterian Church of Ghana became enchanted, let me define the term "deliverance" and discuss its relationship to healing. Deliverance is defined by Catechist Abboah-Offei as setting a person, place, or object free from Satan's control.[1] Deliverance involves deploying the power of the Holy Spirit, particularly with the utterance "in the name of Jesus," or, alternatively, "with the blood of Jesus," in order to free people from any illness, affliction, or other problem caused by Satan. Within highly enchanted forms of Christianity—where illness, health, and healing are influenced by supernatural forces and involve relationships between people and spirits—healing is synonymous with deliverance.[2] Throughout this chapter and subsequent chapters, I will use the

---

1. Abboah-Offei, *Church Leaders Training Manual*, 6–7.
2. As Candy Gunther Brown notes, for most Pentecostals and Charismatics worldwide, divine healing is closely connected with deliverance from demonic oppression

terms "healing," "deliverance," as well as "healing and deliverance" as analogues referring to the same phenomena.

Political, Economic, and Biomedical Decline in Ghana

The context of the institutionalization of deliverance practices within the Presbyterian Church of Ghana was one of political, economic, and biomedical decline. Between 1966, when Ghana's first president Kwame Nkrumah was overthrown by a military coup, and 2000, when John Kufour was democratically elected president, Ghana experienced a series of coups and an alternation between civilian and military regimes.

Over this period, the economy in Ghana suffered considerably, including the once lucrative cocoa industry. By comparison, at independence in 1957, the per capita income of Ghana was on par with Hong Kong, Singapore, Malaysia, and South Korea at about $400. By 2000, Ghana's per capita income had decreased to around $370, while the four Asian polities ranged from $18,000 (South Korea) to $32,000 (Singapore).[3] While 40 percent of the population was living in poverty by the late 1990s, the per capita income of Ghana had increased 3.7 percent between 2003 and 2008 and there are signs of an economic recovery corresponding with political stability.[4] This being said, the 2008 per capita income for Ghana was the same as the peak for the previous thirty-five years, set in 1974.[5] And under these conditions, women suffered so extensively that researchers have begun to refer to the "feminization of poverty" in Ghana, an issue I will explore more fully in chapter 6.[6]

The massive political and economic decline in Ghana between 1960 and 2000 did cause a great amount of suffering, during which many people turned to religious institutions for all sorts of material and spiritual help, including their health. During this period, biomedical healthcare became relatively unavailable to the masses when the government—coerced by structural adjustment programs—introduced user fees for healthcare. Referred to colloquially as the "cash-and-carry" system, patients were not cared for, even in the context of medical emergencies, until service fees were paid. Under this system

---

("Introduction," 18).
   3. Gifford, *Ghana's New Christianity*, 11.
   4. Killick, *Development Economics*, 418, 421.
   5. Ibid., 427–29.
   6. Wrigley-Asante, "Men are Poor"; Yeboah, "Urban Poverty."

many people died, including patients who were incapable of making payments because of their injuries and whose relatives or friends were not immediately available to pay the medical fees.[7]

Even with the introduction of Ghana's National Health Insurance Scheme in 2003, between 65 percent and 82 percent of the population remained uninsured by 2010.[8] Many find the annual fee of around nine dollars prohibitive.[9] Some physicians have also denied care to insurance holders because of the long reimbursement delays by insurance providers.[10] And the massive medical brain drain still continues, where trained Ghanaian medical practitioners—both doctors and nurses—emigrate to find higher paying jobs in the global North (discussed in the next chapter).[11]

While some research has drawn a direct relationship between poverty-induced disease and the rise of Christian healing practices,[12] the same argument cannot be made for the Presbyterian Church of Ghana over time. This causative relationship holds true after 1960; however, the same cannot be said for the prior forty years. As I demonstrated in the previous chapter, it was in the great economic boom of the 1920s that Faith Tabernacle—which abstained from using biomedicine—emerged among a literate, wage-earning class of Presbyterians who were relatively wealthy and had access to better-than-average biomedical care in the various Presbyterian medical facilities. In the late colonial period, it was rather the social problems resulting from great wealth, not from economic depravation, that spurred enchanted Christianity in Ghana. What was consistent through these two boom-and-bust periods—the first colonial and the second postcolonial—was the rise of capitalism and the perceived unequal distribution of resources that was based on a more egalitarian sociocultural standard within Akan society. And in this postcolonial bust period,

---

7. "Health Insurance in Ghana," http://www.ghanaweb.com/GhanaHomePage/health/national_health_insurance_scheme.php.

8. By 2008, the government of Ghana claimed that 54 percent of the population had been registered in the health insurance scheme (Awal, "Ghana," 109).

9. Kotin, "Health Insurance"; Oxfam "Don't Copy."

10. Kotin, "Health Insurance."

11. For instance, adjacent to Abboah-Offei's home in Akropong are two large houses (by American standards). One is owned by Abboah-Offei's cousin—a physician in London—while the second is owned by a group of sisters, who all work as nurses in New York City.

12. See Chestnut, *Born Again*.

from 1960 to the end of the century, deliverance expanded in Ghana among all types of churches.[13]

Enchanted Transformations of the Presbyterian Church of Ghana

By the early 1960s, the Presbyterian Church of Ghana decided to formally institute healing practices in the church because of three reasons. First and primarily, the Presbyterian Church of Ghana was reacting to the massive exodus of church members to enchanted churches. Second, the creation of healing groups among those who stayed members of the Presbyterian Church of Ghana, while typically interacting with the new Pentecostal churches, contributed to the institutionalization of healing practices within the Presbyterian Church of Ghana.[14] Third, the international discussions that were ensuing within the World Council of Churches about healing and deliverance in the mainline churches influenced the Presbyterian Church of Ghana.

First, a feature of Ghanaian Christianity between 1920 and 1960—described in the previous chapter—was the emergence of enchanted Christianity, such as Faith Tabernacle and the Church of Pentecost, which incorporated healing into their ritual repertoires. These new churches developed a type of worship that involved healing and the deliverance of malevolent spirits in response to the insecurities and social changes brought on by the cocoa industry and the massive intrastate migration. Many of these new churches were led by and included former Presbyterians, but their great successes influenced even those who remained in the Presbyterian Church of Ghana.

Second, before the 1960s, there were instances of Presbyterians, typically influenced by new Pentecostal churches, beginning to incorporate healing practices into the Presbyterian Church of Ghana.[15] Yaa Abram from Akyem-Awasa (Eastern Region) began praying for and healing the sick while still a member of the Presbyterian Church

---

13. Asamoah-Gyadu, *African Charismatics*; Gifford, *Ghana's New Christianity*; Omenyo, *Pentecost Outside of Pentecostalism*.

14. In fact, research conducted by the Presbyterian Church in the early 1960s showed that among mainline church members Presbyterians were most likely to visit healing-centered churches or prophets (Presbyterian Church of Ghana, "Appendix F," 41–54).

15. Omenyo, *Pentecost Outside of Pentecostalism*, 132–38.

of Ghana in the 1930s. In 1938, a prayer (and presumably healing) group began in the Ramseyer Presbyterian Church in Kumasi, which met with Assemblies of God members for a time. Beginning in 1946, the catechist J. J. Manteaw began to have prayer meetings with the local Apostolic church, where he received the gifts of healing, and he later formed a prayer and healing group in Bechem, in the Brong Ahafo Region.[16] In 1949, the Presbyterian teacher Sakyi-Addo began worshipping with the local Apostolic Church congregation in Agogo-Ashanti, after which he began a small prayer and healing group. The Presbyterian Pastor M. O. Beeko also prayed for and healed many sick people in Odumase, Swedru, Asamankese, and Nkawkaw (Eastern Region) beginning in 1957.[17]

Third, in August 1963 the Presbyterian Church met to discuss the World Council of Churches' New Delhi and Ibadan report, "The Holy Spirit and the Christian Community," which addressed, among other things, the role of healing in the church. The report to the synod of the Presbyterian Church of Ghana resulting from this meeting, "What Has Happened to Our Prayer Services?," argued for the restoration of the New Testament ministry of healing through prayer within congregations of the Presbyterian Church of Ghana.[18] The committee also claimed that Christian healing should never be isolated from medical treatment, which God can use as a means to heal: a directive for Presbyterians to follow the path of the Church of Pentecost, not Faith Tabernacle and Christ Apostolic Church, which abstained from using biomedicine.[19]

Origins of the Bible Study and Prayer Group

Many individual Presbyterians played roles in establishing healing practices within the Presbyterian Church of Ghana. None, however,

---

16. Manteaw was later stationed by the church at the Presbyterian Church's Agogo hospital as the lay chaplain (Omenyo, "New Wine," 242).

17. Beeko, *Trail Blazers*, 47–51.

18. Omenyo, *Pentecost Outside of Pentecostalism*, 130.

19. While Faith Tabernacle continues to abstain from biomedicine, the Christ Apostolic Church in Ghana does not. Sometime between 1959 and 1964, the Christ Apostolic Church in Ghana began incorporating biomedicine into its healing practices. Interview with Christ Apostolic Church Pastor Samuel Addai-Kusi, Accra, February, 18, 2007.

were more influential than Rev. T. A. Kumi, who helped to form the Bible Study and Prayer Group (BSPG), the primary subchurch organization devoted to healing and deliverance. Kumi's interest in deliverance was in part developed by the Presbyterian Church of Ghana's leadership, which awarded Kumi a two-year scholarship from 1960 to 1962 to study the charismatic revival—the incorporation of Pentecostal practices such as deliverance into mainline churches—within the Presbyterian Church in Scotland in order to introduce these practices into the Presbyterian Church of Ghana.[20] While in Glasgow, Kumi began working with Rev. Clarence Finlayson, founder of the Christian Fellowship of Healing in Scotland, and he later affiliated with the International Order of Saint Luke the Physician (San Diego) and the World Healing Crusade (Blackpool, England).

Kumi's promotion of healing and deliverance upon his return to Ghana directly led to the creation of the BSPG. When Kumi returned in 1962, he became the warden of the Presbyterian Church of Ghana's Ramseyer Retreat Center in Abetifi. By 1963, Kumi was working closely with James McKeown and other leaders of the Church of Pentecost, who were regularly holding church conventions at the Ramseyer Retreat Center.[21] Kumi also began giving seminars on healing and deliverance at the Ramseyer Retreat Center between 1962 and 1965. Many prayer groups in the Presbyterian Church of Ghana were formed as a result of Kumi's seminars. These prayer groups, particularly from Brong Ahafo, Asanti, Kwahu, and Akyem, were formally organized into the BSPG in March 1965, with twenty-one groups and six hundred members.[22] Included in the aims and objectives of the BSPG was "to use prayer as an effective means of enduring and relieving pains, suffering, distress and want."[23]

In the creation of the BSPG, the Presbyterian leadership suggested that perhaps they had focused too much attention on biomedicine and the creation of hospitals, instead of spiritual healing. In these days, the church leadership argued, when modern life breaks kinship ties and thrusts people into great insecurity, the care of individuals must be addressed.[24] Traditional "pastoral care" was not enough;

---

20. Omenyo, *Pentecost Outside of Pentecostalism*, 138.
21. Kumi, "Appendix I," 51.
22. Presbyterian Church of Ghana, "Appendix F," 47.
23. Omenyo, *Pentecost Outside of Pentecostalism*, 142.
24. Presbyterian Church of Ghana, "Appendix F," 51.

prayer fellowships needed to be formed to care for the needs of the various congregations. Qualifying this impetus was the concern to create a standardized set of healing practices amid great diversity within the existent prayer groups. This was the mandate for creating the BSPG.

By 1966, the BSPG was formally recognized as a subchurch organization in the Presbyterian Church of Ghana. Various teams emerged within the BSPG, such as a deliverance team for treating general health as well as cases of misfortune determined to be a result of Satan or demons.[25] Other teams established within the BSPG were counseling teams and a crusaders team, which specialized in evangelism and church planting.[26] These three activities—healing and deliverance, counseling, and evangelism—are quite interconnected within the BSPG and are frequently undertaken by all members within any given BSPG.

Two parachurch organizations, the Scripture Union and the Sudan Interior Mission, also contributed to the rapid growth of the BSPG in Ghana. The BSPG was composed of many members who were also associated with Scripture Union. As one Presbyterian pastor argued, the "Bible Study and Prayer Group flourished because Scripture Union members were often in leadership positions."[27] The Sudan Interior Mission also contributed to the BSPG's proliferation through broadcasting Christian radio programs and distributing Christian literature.

Scripture Union and Sudan Interior Mission

Scripture Union was formed as a nondenominational Christian organization aimed at teaching schoolchildren about the Bible in the United Kingdom in the 1860s, but soon became internationalized by

---

25. These deliverance teams are alternately called prayer warriors, prayer bands, or prayer teams. In the Presbyterian Church, the BSPG made a smooth transition over time to becoming a mainstream organization within the Presbyterian Church of Ghana. Alternatively, in the Evangelical Presbyterian Church in Eweland, members who supported the charismatic movement broke away from the church to found the White Cross Society in 1960 and The Lord's Pentecostal Church (Agbelengor) in 1964 (Omenyo, *Pentecost Outside of Pentecostalism*, 174–76). See Meyer for a more detailed discussion of the emergence of the Lord's Pentecostal Church (Agbelengor) in Peki (*Translating the Devil*, 112–40).

26. Omenyo, *Pentecost Outside of Pentecostalism*, 147–48.

27. Rev. Appiah as quoted in Barker and Boadi-Siaw, *Changed by the Word*, 154.

the end of the century. Scripture Union literature circulated wherever British missionaries were stationed. By 1890, Scripture Union's worldwide circulation had reached 470,000, and its materials were translated into twenty-eight languages.[28] In 1892, the first Scripture Union group met in Accra.

By 1965, Scripture Union Ghana became autonomous after the British organization decentralized.[29] The production of literature—which by the 1960s was aimed at both adults and children—became a hallmark of Scripture Union's evangelical program in Ghana, as it had been elsewhere since the late nineteenth century. In 1965, Scripture Union helped to launch Africa Christian Press to publish Christian books written locally by African authors.[30] Much of the literature soon began to focus on deliverance.

Scripture Union Ghana became a leading organization in advancing deliverance theology and practice within school groups and town fellowships. Scripture Union became the main parachurch organization operating in Ghana's secondary schools by the late 1950s. Graduates with Scripture Union backgrounds had, by the late 1960s and early 1970s, formed groups in universities as well as town fellowships.[31] Some of the members of Scripture Union town fellowships introduced deliverance into the prayer groups of their churches, including the Presbyterian Church of Ghana.[32] Scripture Union's deliverance ministry, led by a team called the Prayer Warriors, emerged alongside the BSPG with much cross-fertilization between the two.

Many of the Scripture Union Prayer Warrior leaders had much interaction with the BSPG, including the founder of Scripture Union Prayer Warriors, Edward Okyere. In September 1973, Edward Okyere, a Presbyterian teacher in Ashanti-Mampong, was appointed Scripture Union traveling secretary for the Asante and Brong-Ahafo Regions.[33] Okyere felt anxious about his new job and decided to invite friends to

---

28. Sylvester, *God's Word*, 34.
29. Ibid., 225.
30. Barker and Boadi-Siaw, *Changed by the Word*, 194–95.
31. Sylvester notes that by 1972, 60 percent of the students at the Trinity Seminary—the main protestant theological college in Ghana—had Scripture Union backgrounds (*God's Word*, 222).
32. Omenyo, *Pentecost Outside of Pentecostalism*, 96.
33. Okyere was one of the first six Ghanaian Scripture Union staff members (Barker and Boadi-Siaw, *Changed by the Word*, 217).

a secluded cave where they could pray quietly for his new job and for the welfare of the nation.[34] After Yaw Frimpong-Manso suggested the site, Okyere invited members of the Sekyeredumasi Scripture Union town fellowship to join him for seven days of fasting and prayer in April 1974.

This week of fasting, prayer, and deliverance became an annual event known as the Warriors' Annual Retreat (WAR). A variety of healing miracles occurred over the first twenty years of WAR: barren women became pregnant, the blind were given sight, demons causing epilepsy were cast out, witches were delivered, a girl was raised from the dead, the insane were healed, and all types of diseases were cured.[35] The WAR retreat in 1974 included 29 participants. By 1992 there were 1,244 people from 54 denominations and in 2003 participants numbered 2,830 from over 50 Christian organizations.[36]

Scripture Union's WAR expanded by 1984 to include an annual deliverance workshop to provide formal instruction in healing and deliverance practices to prayer teams that were developing in all the Presbyterian Church's BSPGs as well as other mainline churches. In 1984, the first workshop was held at the Kumasi Catholic Conference Center with 60 participants. By 2003, there were 1,108 participants from 15 ministries and churches.[37]

Besides Scripture Union, another international parachurch organization, the Sudan Interior Mission, influenced the deliverance movement within the Presbyterian Church of Ghana through Christian radio programming (ELWA) and a Christian bookstore (Challenge). Two Canadians and an American founded Sudan Interior Mission as an evangelical Christian mission to Africa in 1893. By 1952, Sudan Interior Mission, in collaboration with the West African Broadcasting Association, took over Radio ELWA based in Monrovia, Liberia. ELWA, an acronym that stands for Eternal Love Winning Africa, began broadcasting in January 1954.[38] From the mid-1950s through the 1980s, ELWA was broadcasting to Ghana in English and local languages, featuring evangelical and fundamentalist Christian

---

34. Barker and Boadi-Siaw, *Changed by the Word*, 90–91.
35. Kumah, *Twenty Years*, 15.
36. Ibid., 13; Barker and Boadi-Siaw, *Changed by the Word*, 91–92.
37. Scripture Union Prayer Warriors, *Twentieth Anniversary*, 7.
38. A brief history of Radio ELWA can be found on its website, www.elwaministries.org, accessed April 9, 2013.

programming.[39] After he became born-again in 1965, Edward Okyere remembers being specifically influenced by ELWA radio, which aired popular Billy Graham programs.[40]

Another organization established by Sudan Interior Mission that strongly influenced the deliverance practices within the Presbyterian Church of Ghana was Challenge Enterprises, an international parachurch organization established in Ghana in 1956, and later managed exclusively by Ghanaians after 1975. In particular, Challenge runs a bookshop, which is Ghana's largest distributer of Evangelical Christian literature. Challenge sells Evangelical Christian literature—from Europe, North America, and West Africa—at discounted prices. Also, Challenge offers a deliverance workshop similar to Scripture Union in the form of the All Pastors Annual Conference, which was inaugurated in 1987, where deliverance literature is also sold at reduced prices.[41]

The deliverance literature disseminated by Challenge Enterprises bookstore to leaders within the BSPG influenced the movement significantly. Of particular importance was the distribution of the Nigerian Emmanuel Eni's book *Delivered from the Powers of Darkness*, published by Scripture Union Nigeria, which became a best-seller in Ghana by 1988.[42] And there was cross-fertilization between Challenge bookstore, the BSPG, and Scripture Union. For instance, the first fulltime secretary of the BSPG, Kwasi Asenso, was a leading Scripture Union member who was employed at Challenge Enterprises bookstore at the time of his appointment in 1976.[43]

These two parachurch organizations—the Scripture Union and Sudan Interior Mission—helped to introduce deliverance practices into the Presbyterian Church via the BSPG. And the Presbyterian Church was the first mainline church in Ghana to accept and support the charismatic movement—particularly healing and deliverance—within its framework.[44] These deliverance practices helped the Presbyterian Church of Ghana to grow significantly, which was another mandate of the BSPG: evangelism and church growth. In this mandate, the Presbyterian Church of Ghana was very successful. During a

---

39. Debrunner, *History of Christianity*, 349.
40. Author interview with Edward Okyere, Akropong, Ghana, February 11, 2007.
41. Asamoah-Gyadu, *African Charismatics*, 170; Onyinah, *Akan Witchcraft*, 225.
42. Asamoah-Gyadu, *African Charismatics*, 171.
43. Barker and Boadi-Siaw, *Changed by the Word*, 154.
44. Omenyo, *Pentecost Outside of Pentecostalism*, 146.

Figure 3.1. Challenge Enterprise's bookshop signboard, Accra, February 17, 2007. Photograph by the author.

five-year span between 1987 and 1991, the Ghana Evangelism Committee, which conducts surveys of Christian membership in Ghana based on Sunday service attendance, reported a 17 percent growth in the Presbyterian Church of Ghana. The only other mainline church to experience growth rather than a decline over this period was the Methodist Church, which grew only 2 percent.[45]

The institutionalization of deliverance practices within the Presbyterian Church of Ghana was furthered in the mid-1990s with the establishment of the first charismatic Presbyterian Church, Grace Presbyterian, which serves as a deliverance center: public deliverance services are preformed and private consultations are offered to the suffering.[46] Grace Church was established by Catechist Ebene-

---

45. Gifford, *African Christianity*, 62.

46. The Presbyterian Church of Ghana has other charismatic Presbyterian churches, besides Grace, that serve as deliverance centers. By 2007, five other charismatic Presbyterian churches existed in Ghana: Atibie (Kwahu), led by Rev. Nana Ntim Gyakari; Jejemireja (Brong Ahafo), led by Brother Daniel Ansu; Gilbal Prayer

zer Abboah-Offei, a leader in the BSPG and Scripture Union, who was significantly influenced by literature from Challenge Enterprises bookstore.

Catechist Ebenezer Abboah-Offei, Deliverance Practitioner

On June 23, 1957, Ebenezer Abboah-Offei was born in Akropong into one of the oldest Christian families in the country.[47] One of his ancestors, David Asante, was the first ordained Ghanaian pastor of the Basel Mission; he trained at the Basel Seminary in greater Württemberg between 1857 and 1862.[48] Abboah-Offei's father had been a cocoa farmer and cocoa broker in Suhum, and one of many Akwapim Presbyterians who purchased land for cocoa farming in Akyem Abuakwa. Abboah-Offei lived in Suhum as a small boy until his father's death.[49] Abboah-Offei, aged four or five, then moved back to Akropong to attend primary school, after which he moved to Asamankese, where his family managed a 300-acre palm oil plantation (used initially for cocoa), for secondary school. Abboah-Offei spent the next twelve years living in Asamankese and working on his family's palm oil plantation as well as in his family's soap factory. While living in Asamankese in 1975, he became born-again as a result of joining Scripture Union in secondary school.

Abboah-Offei, like many other young Presbyterians whose families were involved in cocoa production, was drawn to more enchanted forms of Christianity. While living in Asamankese—the birthplace of Pentecostalism in Ghana—Abboah-Offei visited many churches with robust healing practices, such as the Church of Pentecost, the Christ Apostolic Church, and the African Faith Tabernacle.[50] At age

---

Center in Boma (Brong Ahafo); Miremano (Brong Ahafo), led by Brother Andrew Kye; and Chirapatre in Kumasi (Asante), led by Brother Fred Darko.

47. Abboah-Offei is a member of the Nkatia lineage, which is one of the three royal lineages in Akropong. Akropong, the capital of the Akwapim polity (*Okuapeman*), was formed in 1733 by three founding women, each of whom formed one of the three royal lineages. The chief of Akropong (*Okuapehene*) is a rotating office between each of these three lineages.

48. Debrunner, *History of Christianity*, 305.

49. Author interview with Ebenezer Abboah-Offei, Akropong, November 26, 2007.

50. The African Faith Tabernacle, led by the successful cocoa farmer James Kwame Nkansah in Anyinam, split from Faith Tabernacle in Philadelphia in 1953

twenty, Abboah-Offei joined two Christian student organizations: the Asamankese Christian Fellowship (where he became president) and Joyful Way Incorporated, an evangelistic youth music team that helped to modernize Christian music in Ghana.[51] From 1979 to 1986, Abboah-Offei was president of the Scripture Union town fellowship in Asamankese.

In the 1980s, Abboa-Offei's career in agriculture began to blossom. In 1985, Abboah-Offei graduated from the University of Science and Technology in Kumasi with an agricultural degree and worked from 1985 to 1987 as the plantation manager of his family's 300-acre palm oil plantation. Simultaneously, he managed a government-owned palm oil plantation located on 10,000 acres of land 20 kilometers outside of Asamankese, which employed over eight hundred workers. Abboah-Offei's commitment to his faith—he was still a member of the Presbyterian Church—increased as he became more successful. As plantation manager, Abboah-Offei would show Christian films, such as the *Ten Commandments*, for the laborers during nonwork hours. These films were purchased from Challenge Enterprises bookstore in Accra.

While his spiritual and professional life was thriving in Asamankese, God spoke to Abboah-Offei one day and directed him to move home to Akropong. Soon after his marriage in September 1987 to Faustina Martinson, the organizing secretary of the Asamankese Christian Fellowship, Abboah-Offei followed God's directive and moved with his wife to Akropong. He began to teach agriculture at the Okuapemman secondary school that year.

In order to supplement his income as an agricultural teacher, Abboah-Offei started a small poultry farm in Akropong. Within a year he had 3,000 chickens. At this time, Abboah-Offei, who was not yet a deliverance practitioner, noticed an ability to heal his sick animals through prayer.[52] Abboah-Offei's success in praying for the health of his animals became more apparent one day when his chickens

---

(Baeta, *Prophetism in Ghana*, 114). In early 1952, one year before parting ways with Faith Tabernacle, Nkansah claimed to have two thousand people attending his services in Anyinam. FTC-CN, James K. Nkansah to Walter Troutman, January 16, 1952.

51. Asamaoh-Gyadu, *African Charismatics*, 107.

52. Lest the practice of healing animals through prayer sound singular, it was very common for Johan Christoph Blumhardt, while pastoring in the farming town of Mottlingen in the mid-nineteenth century, to pray successfully for his congregants' sick animals (Zuendel, *Awakening*, 100).

were accidentally given poisoned feed. Abboah-Offei had unknowingly bought chicken feed containing the toxic chemical aflatoxin. Immediately after giving this feed to his chickens, 157 collapsed dead. Abboah-Offei quickly began praying, asking God to save the other chickens. After praying for a while with the chickens, Abboah-Offei asked his assistant to fetch him a bucket of water. Abboah-Offei took the water and sprinkled it on the living chickens, telling them that through the blood of Jesus, baptism brings life from death. Not another bird died after the prayers and baptism.

After this event, Abboah-Offei was determined to apply his gift of prayer to the welfare of people, not animals. As a result, Abboah-Offei joined the BSPG at Akropong's Christ Presbyterian Church, the oldest continuously attended Presbyterian church in Ghana, established in 1835 by the missionary Andreas Riis. By 1989, Abboah-Offei was appointed the BSPG president at Christ Church and soon became an elder, a rare appointment for a man not yet thirty years old. Many of the older leaders in the church disagreed with Abboah-Offei's promotion of deliverance practices, claiming they were unbiblical. But the BSPG at Christ Church continued growing. During the late 1980s, Abboah-Offei was reading deliverance literature voraciously, particularly publications sold at Challenge Enterprises bookstore.

As the BSPG began to grow within Christ Church, with more and more people flocking to deliverance services, the leadership at Christ suggested that Abboah-Offei found a charismatic Presbyterian Church in Akropong because of tensions within Christ Church's congregation; many did not support the charismatic movement and the church leaders were concerned about congregational instability.[53] Abboah-Offei's vision for this type of institution was particularly shaped by similar deliverance centers that were established by the Church of Pentecost. In particular, during the mid-1990s, Abboah-Offei was influenced by the Bethel Prayer Camp in Sunyani, which was led by the Church of Pentecost's Owusu Tabiri.[54] The patronage of the Church of Pentecost prayer camps by mainline church

---

53. Christ Church maintained a strong BSPG after Abboah-Offei left. By early 2007, the Rev. George Opare-Kwapong, who received a master of theology degree at Princeton University, was the pastor of Christ Church. Opare-Kwapong was the very first member of Abboah-Offei's Grace Team and is an active supporter of Christ Church's BSPG. Opare-Kwapong continues to be a member of the Grace Team.

54. Abboah-Offei discussed Tabiri's focus on breaking with ancestral spirits during a lecture at the deliverance workshop in New York on November 26, 2007.

members such as Abboah-Offei was so significant that the mainline churches began establishing their own prayer camps or deliverance centers, such as Grace Presbyterian Church in Akropong.[55]

While the establishment of Grace Presbyterian Church placed Abboah-Offei at the center of Presbyterian Church of Ghana's deliverance movement, his kinship network also has influenced this process. T. A. Kumi, the primary person responsible for establishing the BSPG, is the great-uncle of Abboah-Offei, while Edward Okyere, who inaugurated Scripture Union's WAR, is Abboah-Offei's cousin.[56] M. K. Yeboah, the former Presbyterian who became the Church of Pentecost chairman from 1988 through 1998, is Okyere's father.[57] Finally, Samuel Asare, who is Abboah-Offei's assistant at Grace, is the godson of Faustina Abboah-Offei, the wife of Ebenezer Abboah-Offei. Healing and deliverance has become a family affair in Abboah-Offei's extended Akropong family, although he has become well-known in his own right through establishing and managing Grace Presbyterian Church, which I describe in more detail in the following section.

Grace Presbyterian Church

Grace Presbyterian Church was founded in March 1996 by Abboah-Offei—with the help of his assistant Samuel Asare and with the encouragement of his wife Faustina Abboah-Offei—on the grounds of the Presbyterian girl's primary school in Akropong. By 1999, Abboah-Offei had quit his teaching job to devote himself full time to the Presbyterian Church's deliverance ministry and to Grace Church.[58] In 2001, Abboah-Offei was consecrated as a catechist or lay pastor in the Presbyterian Church of Ghana.

After Abboah-Offei had sold all his chickens, he used the wood from the chicken coups to build Grace's first sanctuary. In time, sufficient money was raised through tithes and offerings from Abboah-Offei's deliverance services to begin construction of a much larger

---

55. Onyinah, *Pentecost Outside of Pentecostalism*, 255.

56. Kumi married Abboah-Offei's grandfather's sister.

57. Yeboah is not, however, Abboah-Offei's uncle. Yeboah married the wife of Abboah-Offei's uncle, after he had passed away.

58. But before he stepped down from teaching at Okuapemman, Abboah-Offei created a prayer team to pray for sick students and deliver them from the demons causing widespread illness during exam time (Barker and Boadi-Siaw, *Changed by the Word*, 128).

concrete sanctuary for Grace Presbyterian Church to replace the original wooden structure as the main venue for public services. By 2007, the basement of the larger sanctuary was completed and used to hold services. The original wooden sanctuary is now used for individual consultation for the afflicted. Plans are underway to raise funds to complete the main sanctuary. Like many structures in Ghana, it is being built piecemeal as the necessary funds are secured. Also in 2007, several smaller stucco buildings were erected on the property around Grace. One building is used by Abboah-Offei as his office, while another houses the patient files.

A typical week at Grace includes individual consultation available on Mondays, Tuesdays, and Thursdays. Usually, 25 to 50 people show up per day between the working hours of 8 am and 3 or 4 pm. Wednesdays and Saturdays are reserved for public deliverance services from 1 to 5 pm, after which individual consultation is available. Wednesday service tends to be smaller, drawing 150 to 300 people, while Saturday service is attended by 300 to 500 people (see fig. 3.2). Friday is the day of rest, prayer, and meditation for Abboah-Offei, while Sunday is reserved for regular church service (as opposed to specific deliverance service). One weekend a month from February through November Abboah-Offei teaches his deliverance course at Grace.

As of 2007, Grace Church employed over sixty deliverance workers, referred to as the Grace Deliverance Team or more simply the Grace Team, which is the BSPG of Grace Church. All members of the Grace Team have successfully passed Abboah-Offei's deliverance course. Many of these people are retirees who are living off a pension and have free time to participate in their church. Some of the workers have other full-time jobs and come to volunteer at Grace whenever they are available. Still other team members are people who cannot find other work and are paid a small salary by Abboah-Offei.

While this deliverance system practiced at Grace is exclusive of and demonizes Akan spiritual healing practitioners, it includes biomedicine, as the 1963 report of the Presbyterian Church of Ghana specified. One important aspect of this healing system within the Presbyterian Church of Ghana is its incorporation of biomedicine and psychiatry, as opposed to Faith Tabernacle, which abstains from all other forms of healing. There is one volunteer psychiatrist who assists in diagnosing and treating various disorders at Grace. Also, several of the retired Grace Team members are former nurses or doctors.

Figure 3.2. Saturday service at Grace Presbyterian Church, Akropong, February 17, 2007. Photograph by the author.

The Grace Team frequently accepts people from and sends people to the nearby Tetteh Quashie Memorial Hospital in Mampong, which is typically facilitated by the staff who work or have worked for both institutions.[59]

Deliverance Services and Workshops at Grace Presbyterian Church

There are three primary activities that dominate Grace Presbyterian Church: public deliverance services, deliverance workshops, and individual consultation. I will describe the first two in this section. At public deliverance services, Abboah-Offei either dresses in a dark suit and tie or traditional Ghanaian cloth when preaching. His sermons are read from notes on his laptop, while PowerPoint presentations

---

59. For a slightly parallel case, see Langwick's study of nurses in a Tanzanian hospital who help incorporate indigenous healing practices both within and outside the hospital (Langwick, "Articulate(d) Bodies").

are projected onto a large screen for the congregation to follow. Services are predominantly performed in Twi, although some English is used. The services typically begin with worship and prayer, which includes music, dancing, and praying. This is followed by Abboah-Offei's sermon, which is organized around a central theme focusing on deliverance.

During deliverance sermons, spirit possession often occurs spontaneously. By spirit possession, I am referring to executive possession, which entails spiritual agent(s) taking over a host's executive control or replacing a host's mind. Furthermore, in the context of deliverance, spirit possession entails executive possession by malevolent spirits and not the Holy Spirit. When possession occurs during sermons at Grace Church, various Grace Team members care for and subdue the possessed person. At other times, Abboah-Offei will pick people out of the audience who have a particular disorder as they are revealed to him (by God) during a service. He will call them forward, lay hands on them, pray over them, and frequently they will become possessed. Illness experienced through possession is a topic I will discuss in great detail in chapter 6.

Usually time is given for testimonies after the sermon is delivered. People give thanks for health and success in their lives, such as healing from afflictions, success in school, at work, or in a relationship, reproductive success, or finding protection from malevolent spiritual forces. Many times after the sermon and testimonies are completed, a prayer line will be called, where people with particular afflictions or misfortunes will be called forward and prayed for. This too can result in possession. Prayer lines are followed by offerings, more prayers, and, finally, the benediction, where Grace Team members continue to lay hands on people and pray with individuals and groups of people with various problems.

In July 2000, Abboah-Offei began offering a deliverance workshop at Grace Church. Abboah-Offei's two-year course is offered one weekend a month from February through November, for ten months total. It is offered only to prayer teams, not individuals. For a prayer team to pass the course the members must attend for two years or twenty total weekends. In the course of a weekend, teaching takes place Friday evening and all day Saturday. Abboah-Offei estimated that at the start of the program's seventh year in February 2007, approximately three hundred people had graduated from his course. Abboah-Offei trains prayer teams from the mainline churches, such as Methodists, Presbyterians, and Catholics, as well as Neo-Pentecostal prayer teams. They have yet to train any teams from the classical Pentecostal churches such as the Church of Pentecost.

Before leading his own deliverance workshop, Abboah-Offei had a significant amount of experience attending and later leading the Scripture Union deliverance workshop.[60] Abboah-Offei first participated in the Scripture Union deliverance workshop as early as 1987; by 2000 he was teaching deliverance procedures in the Scripture Union workshop, and in December 2006, Abboah-Offei gave the keynote address (called theme address). One primary difference between the Scripture Union course and the one offered by Abboah-Offei is the greater body of research from both primary and secondary sources incorporated into the catechist's course and course materials. Primary source material includes research conducted among Akan religious and healing practitioners by Abboah-Offei's Grace Team, while his secondary source material includes a plethora of deliverance authors such as Rebecca Brown, Mark Bubeck, Abraham Chigbundu, Peter Horrobin, Harold Horton, Samuel O. Onwona, Francis MacNutt, Ed Murphy, and Dean Sherman. Besides deliverance authors, other secondary sources include anthropological monographs on the occult in Africa. Abboah-Offei was particularly influenced in thinking about witchcraft as a pan-African phenomenon by E. E. Evans-Pritchard's research on witchcraft among the Azande.[61]

Individual Consultation at Grace Presbyterian Church

Besides deliverance services and workshops, Grace Church also offers individual consultation for the afflicted. Individual consultation always began with the sufferer filling out a deliverance questionnaire used to diagnose his or her affliction. The questionnaire administered by Abboah-Offei, which is the same one used by Scripture Union, was adapted from one used by the Anglican Church of Singapore, which

---

60. Besides Scripture Union deliverance workshops, Abboah-Offei was influenced by Peter Horrobin and Abraham Chigbundu, who both teach deliverance courses in the United Kingdom and Nigeria, respectively.

61. Abboah-Offei discussed Evans-Pritchard's research in detail during a deliverance course taught in Philadelphia in April 2006.

Abboah-Offei procured through international Scripture Union networks. Abboah-Offei's deliverance questionnaire has seven sections that ask particular questions about a person's past (see appendix). Section one asks for personal information, such as name, clan, occupation, and marital status. Section two asks the counselee to describe the affliction, its duration, and any previous actions taken to alleviate it. Section three asks questions pertaining to family background, such as whether one's parents had worshipped in Aladura churches, whether one's family members had visited "native doctors" or ancestral shrines, and whether there are ancestral stools in the family or curses laid upon any family members. Section four asks questions similar to section three, but with respect to the afflicted individual, not to his or her family. Section five asks about unusual phenomena experienced such as seeing hallucinations, hearing strange voices, or having voices inside one's head. Section six asks questions about antisocial behavior, such as excessive hatred, anger, or sexual disorders. Finally, section seven asks about unusual phenomena experienced in dreams, such as being abused, losing money, or talking with dead friends or family. Answers to sections one and three through seven will give an indication of what is causing the affliction indicated in section two, if the individual's problem has a spiritual cause.

At Grace Church, in the context of individual consultation, the afflicted are referred to as counselees, while the Grace Team members are referred to as counselors. In filling out the questionnaire, counselees are assisted by a group of women from the Grace Team if it is their first time at Grace. If not, they fill out the forms themselves unless they request help. This takes place at a table set up in the new sanctuary. With new counselees, the Grace Team members explain each question and encourage them to answer all questions accurately and honestly.

Questionnaires used by the Grace Team are individually marked with a Grace Team stamp. This is done in order to avoid people bringing in fraudulent forms for individuals not seen by or prayed for by the first set of counselors, who assist in filling out the forms and praying for the counselees. Each questionnaire is also marked with a unique number according to the number of visitors to date in that calendar year: for instance, 304/10 is the 304th counselee since January 2010. Counselees are given a card that lists their name, hometown, and the number corresponding to their deliverance questionnaire. If an individual comes for future counseling, the Grace

Team will be able to pull his or her file using this cataloging system. The deliverance forms are stored in archives on the premises, and one of the Grace Team members is assigned the task of maintaining and retrieving these forms.

Before being sent with their form to an individual counselor, counselees are made to affirm or accept Christ in their lives. This is a necessary component of deliverance, and a transitional phase between filling out the questionnaire and meeting with a counselor. Anywhere from one to three of the women who assisted in filling out the questionnaire will bring the counselee to a corner of the sanctuary, lay hands on him or her, and ask if the counselee is born-again (see fig. 3.3). All counselees need to have already accepted Christ as their personal savior, accepted Jesus during this interaction, or agreed to have a covenanted fellowship with Him. Without this, a counselor will not see a counselee.

After this affirmation, Grace Team members will pray with the counselee against general afflicting agents, particularly to cast out demons in their families, destroy family altars, and combat witchcraft. Then the deliverance counselor will pray for the specific affliction listed in question two of the deliverance questionnaire. Finally, when all these steps have been completed, the counselee is sent from the main sanctuary to a covered waiting area outside the old wooden sanctuary. When a counselor becomes free, he (all are men) calls the counselee to one of four tables set up inside the sanctuary. Each table is equipped with olive oil for anointing and tissue paper for cleaning substances vomited if possession occurs during the course of the consultation.

While seated in a chair at a small table opposite the counselee, a counselor will discuss the questionnaire with the counselee in an attempt to diagnose the problem based on the data collected (see fig. 3.4). Discussion ensues and the counselors take notes directly on the questionnaire, to which counselees are not given access. Because many of the spiritual disorders stem from interpersonal relationships, it is the policy of the Grace Team *not* to reveal the source of the afflicting agent to the sufferer for fear of retribution. Only the counselors are aware of the exact nature of the diagnosis. For example, counselors would reveal that a person's chest pain was caused by a witchcraft attack, but they would not reveal that a counselee's aunt is the person bewitching him or her. If an afflicted person does discover that a living individual is in some way responsible for an illness or misfortune—

Figure 3.3. Counselee with hand raised made to affirm or accept Jesus Christ in her life before consulting a deliverance counselor at Grace Presbyterian Church, Akropong, February 12, 2007. Photograph by the author.

which is very common in sub-Saharan Africa in general[62]—it is the policy at Grace Church that he or she *not* approach the person or seek retribution, but rather pray for that person's forgiveness.

While the completed questionnaire with resultant discussions often lead to a diagnosis, nondiscursive signs are equally important in this therapeutic process. For instance, during one individual consultation,

---

62. See Westerlund, *African Indigenous Religion.*

Figure 3.4. Counselor meeting with counselee at Grace Presbyterian Church, February 12, 2007. Photograph by the author.

Samuel Asare prayed for a woman who began pulling at her ring finger, then pulling all her fingers, and then rubbing her hands together. On her questionnaire, she wrote that she had sex in her dreams with strangers. The sign of pulling her ring finger, which is an index of spiritual marriage, in conjunction with the information about having sex in her dreams with strangers, led to her diagnosis: the woman had a spiritual marriage with a demonic being.

Another diagnostic technique, to determine whether a disorder has a demonic etiology, used during individual consultation as well as during public deliverance services, is diagnosis via therapy.[63] More specifically, a counselor will pray with a counselee against a variety of afflicting agents, one after the other, until possession occurs, which signals the correct afflicting agent. Diagnosis via therapy frequently occurs when completed questionnaires, resultant discussions, and nondiscursive signs do not reveal the afflicting agent.

---

63. See Feierman, "Therapy as a System-in-Action."

Interaction between counselor and counselee in the course of individual consultation takes different forms. Sometimes during the consultation at Grace Church, counselors will only discuss the problems with counselees. At other times prayers are given, and the counselor lays hands on the counselee. Sometimes counselors anoint counselees with olive oil. On other occasions, prayers are given and counselees become possessed; in such instances other Grace Team members will be called on to care for and restrain the possessed.

While consultations are being performed, many Grace Team members participate as intercessors, occupying a section of the new sanctuary and praying for various things. There are usually ten to fifteen intercessors praying at any given time on Mondays, Tuesdays, and Thursdays, when Grace Church is open for individual consultation. If there is an urgent problem, then Abboah-Offei will indicate to the intercessors that a certain affliction needs to be prayed for or that a certain malevolent agent needs to be prayed against. If there is no urgent problem, then the intercessors will pray separately or join in prayer led by a leader who will announce a topic of his or her choice. These topics follow a prayer line, which is a series of designated prayers on a specific topic such as illness, witchcraft, or demons.

Deliverance Outside of Grace Presbyterian Church

Outside of planned deliverance services, many people come directly to Abboah-Offei's home, located one or two kilometers from Grace Church, for consultation or deliverance. People arrive at all hours of the day and night. He meets with and counsels people from all strata of Ghanaian society: from secondary schoolchildren trying to get into university and young women who have fallen into prostitution, to federal judges and successful Accra businessmen. But many in more urgent need—those suffering from sickness or misfortune and those demonically possessed—go directly to Abboah-Offei's home, where they are cared for by him and his wife. Finally, people often stop by Abboah-Offei's house on their way to the Tetteh Quashie Hospital, so that they or their family members can be prayed for before consulting a physician.[64]

---

64. Abboah-Offei remembers one instance in which a couple brought their young daughter, who was unconscious and nearly dead, to his house before taking her to the hospital. Some neighbors got involved and threatened to sue Abboah-Offei if

Other structures used for deliverance purposes are in the process of being erected on land privately owned by Abboah-Offei adjacent to his home. In 2007, the catechist was contemplating building a retreat center on a piece of this land.[65] Adjacent to this proposed site is a tract of land approximately 150 feet by 150 feet, where Abboah-Offei planted palm trees used for outdoor deliverance services. Typically at the outdoor services, groups of intercessors are placed in different areas of the prayer garden. Each group of intercessors prays for one specific matter, such as protection against witchcraft, success at work, or reproductive fertility. People come to the palm garden and pray in an area dedicated to their specific disorder.

Conclusion: Ghana's Enchanted Calvinism

By the first decade of the twenty-first century the Presbyterian Church of Ghana had become enchanted. The Presbyterian Church leadership formally recognized the multiplicity of spiritual entities that cause affliction in the world, as well as the power of the Holy Spirit to heal the suffering. With the creation of the BSPGs as well as Grace Presbyterian Church, there became a multiplication of practitioners—some particularly charismatic like Catechist Ebenezer Abboah-Offei—to manage this therapeutic process and heal the suffering. The leadership of the Presbyterian Church of Ghana in recent years has also facilitated this process by being very pro-deliverance: Yaw Frimpong-Manso, moderator of the Presbyterian Church of Ghana from 2004 to 2010, helped establish Scripture Union's WAR with Edward Okyere in 1974, and the current moderator, Rev. Dr. Emmanuel Marty (2010–present), was the first theologian to teach deliverance seminars at Trinity College, Ghana's mainline seminary, in the late 1990s.[66]

---

the child happened to die because she had not been taken immediately to the hospital. He prayed over the girl: she began shaking and ended up regaining her consciousness and strength before she left his house. Author interview, Ebenezer Abboah-Offei, Akropong, February 15, 2007.

65. The retreat center will be primarily used to house prayer teams attending his deliverance course in Akropong. The model being followed is based on the Presbyterian retreat center in Stony Point, New York, where Abbaoh-Offei teaches his deliverance course annually for the Ghanaian Presbyterian prayer teams in North America.

66. Gifford, *Ghana's New Christianity*, 89.

In this period, from 1960 to the second decade of the twenty-first century, the Presbyterian Church of Ghana became extremely enchanted under social conditions of extreme political and economic deterioration. The declining political economy in Ghana after independence led to the suffering of the masses—and women more than men (see chapter 6)—that could be addressed by the enchanted Christianity now offered within the Presbyterian Church of Ghana. This context of political and economic decline also corresponded to the development of the Neo-Pentecostal (variously called Charismatic or Born-Again) movement in Ghana and also Nigeria.[67] In fact, many of the same institutions and people who were partaking in the Neo-Pentecostal movement were also playing a role in or interacting with people associated with the Presbyterian Church of Ghana, particularly Catechist Ebenezer Abboah-Offei.

At the institutional level, the same parachurch organizations that helped to develop the Presbyterian Church's BSPG—Scripture Union and Challenge Enterprises bookshop—also played a significant role in the Neo-Pentecostal movement. For instance, Paul Gifford cites the success of Emmanuel Eni's *Delivered from the Powers of Darkness*—a Scripture Union publication—as an early example of deliverance literature that became a best-seller in Ghana, influencing Neo-Pentecostals.[68] Scripture Union, too, became the leading parachurch organization within Ghanaian secondary schools, which fostered much deliverance among Neo-Pentecostals.[69] Scripture Union's deliverance workshop—in which Abboah-Offei now plays a leading role—is also given credit in leading to the popularity of deliverance practices among Neo-Pentecostals.[70]

Challenge Enterprises bookshop, through selling books and offering deliverance courses, also played a major role in fostering deliverance practices among Neo-Pentecostal pastors, particularly through their Annual Pastors' Prayer Conference, which began in 1987.[71] Challenge played a similar role within the healing and deliverance movement in the Presbyterian Church of Ghana. Abboah-Offei was inspired by Mark Bubeck, an American deliverance author who sig-

---

67. Gifford, *Ghana's New Christianity*; Marshall, *Political Spiritualities*.
68. Gifford, *Ghana's New Christianity*, 85.
69. Asamoah-Gyadu, *African Charismatics*, 102–4.
70. Ibid., 170.
71. Ibid.

nificantly influenced the deliverance movement within Ghana's Neo-Pentecostal Churches, with whom Abboah-Offei met at Challenge in 1983.[72] And other authors such as Abraham Chigbundu and Rebecca Brown—who influenced deliverance within the Neo-Pentecostal movement[73]—played a similar role within Abboah-Offei's thinking and practice.

Therefore, the enchantment of the Presbyterian Church of Ghana, through the institutionalization of deliverance practices, has parallels with the Neo-Pentecostal movement in Ghanaian Christianity. The same social forces and institutions affected both movements, which I argue have led to the enchantment of Ghanaian Christianity generally, whether mainline, Pentecostal, or Neo-Pentecostal. In the following three chapters, I will recount the establishment and development of enchanted Calvinism among Ghanaian Presbyterians in North America.

---

72. Ibid., 171. Catechist Abboah-Offei met Bubeck at Challenge bookshop during an event cohosted by Scripture Union and Bubeck's publishing company (possibly Moody Press in Chicago). Abboah-Offei fondly remembers conversing with Bubeck, who encouraged him always to be humble and to study scripture intensely. Author interview, Ebenezer Abboah-Offei, Akropong, February 15, 2007.

73. Gifford, *Ghana's New Christianity*, 89.

Part 2

North America

4

The School of Deliverance and the Enchantment of the Ghanaian Presbyterian Churches in North America

This chapter marks a transition in space from Ghana to North America. In chapter 4, as well as chapters 5 and 6, I will explain the rise of religious enchantment among Ghanaian Presbyterians in North America. This chapter focuses specifically on the primary institution that led to the training of deliverance practitioners in North America: the New York deliverance workshop or school of deliverance. Through this deliverance workshop, primarily, the Ghanaian Presbyterian Churches in North America became enchanted: training charismatic healers to combat the multitude of afflicting spirits that cause great suffering in this Ghanaian Presbyterian community.

The narrative of transnational African religion with familiar themes like witchcraft, afflicting spirits, and spirit possession is much more commonly told through the lens of African-derived religion in the New World such as Brazilian Candomble,[1] Haitian Vodou,[2] or Cuban Santeria.[3] Similarly, several ethnographies have focused on the transnational nature of African religions brought to the United States by practitioners from the Caribbean and Latin America to serve adherents in the United States.[4]

---

1. Capone, *Searching for Africa*; Matory, *Black Atlantic Religion*.
2. McAlister, *Rara!*; Ramsey, *Spirits and the Law*. McAlister's book also has a chapter dealing with Haitian immigrant religion in the United States.
3. Clark, *Where Men Are Wives*; Wirtz, *Ritual, Discourse, and Community*.
4. Brown, *Mama Lola*; Pelrez y Mena, *Speaking with the Dead*; Richman, *Migration and Vodou*.

Related African religious traditions still do exist and thrive on the African continent.[5] However, as of 2010, 63 percent of sub-Saharan Africa was Christian and a great majority of the 1.5 million African-born immigrants and their families living in the United States today are Christian.[6] Recent West African Christian immigrants have transplanted their religious communities with all the same creativity and innovation as their Candomble, Santeria, and Vodou counterparts. But beyond a few recent edited books,[7] very little has been written about African-born Christian communities in North America.[8] Even less has been written about West African Christianity in North America focusing on the same phenomena common among other studies of African religions—witchcraft, spiritual affliction, and spirit possession—which some scholars have argued are the defining features of African Christianity.[9] The three chapters that follow will give a detailed ethnographic account of the enchantment of the network of Ghanaian Presbyterian churches in North America.

Ghana's Declining Economy and Transnational Migration

After Ghana achieved independence in 1957 and into the 1960s, large-scale rural to urban migration ensued as the economy declined, particularly within the rural-based cocoa sector. In 1956, cocoa producers were earning the same amount of money per year as an average worker in Accra. By 1964, cocoa producers were earning less than half of what they earned in 1956, while average earnings per year in Accra increased slightly.[10] The excessive taxing of cocoa after 1960, used to finance other development initiatives along with suboptimal incentives and distorted prices, destroyed the cocoa industry in

---

5. See McCaskie, "Akwantemfi."
6. Caps, McCabe, and Fix, *New Streams*; Olupona and Gemignani, "Introduction"; Pew Research Center, *Global Christianity*.
7. Olupona and Gemignani, *African Immigrant Religions*; Ludwig and Asamoah-Gyadu, *African Christian Presence*.
8. But see Biney, *From Africa to America*. Significantly more has been written about African immigrant Christianity in Europe, however. See Daswani, "Transformation and Migration"; Fumanti, "Viruous Citizenship"; Harris, *Yoruba Diaspora*; and Ter Haar, *Halfway to Paradise*.
9. Cox, *Fire from Heaven*; Ter Haar, *How God Became African*.
10. Mikell, *Cocoa and Chaos*, 189.

Ghana.[11] The cocoa industry was so severely mismanaged by the Acheampong government (1972–78) that producers frequently smuggled cocoa over one of Ghana's land boarders with Togo or Côte d'Ivoire in order to get a reasonable price for their produce.[12] Cocoa farming was no longer the lucrative business it once was, compared to other sorts of urban industries and jobs. And women continued to suffer more than men within this declining economy.

With continued economic deterioration—such as low crop prices and high inflation—large-scale emigration took off in the 1970s. Nigeria became the first major site of Ghanaian transnational immigration, which was fueled by the oil boom in the 1970s. The economic conditions continued to deteriorate in Ghana through the 1980s and into the 1990s, owing particularly to the implementation of the IMF/World Bank-sponsored Structural Adjustment Programs (SAPs) in 1983.[13] In that same year, Nigeria expelled over a million Ghanaians within a forty-eight-hour period: 50 percent of these migrants were from the Asante and Eastern Regions.[14] The implementation of SAPs, combined with the massive expulsion of Ghanaians from Nigeria in 1983, marked the beginning of a large-scale emigration of Ghanaians to North America, and to the United States primarily. While Ghana's economy began stabilizing by the turn of the twenty-first century, the pattern of large-scale Ghanaian emigration to United States continued.[15]

Three immigration policies led to large-scale Ghanaian immigration to the United States. First, the 1965 US Immigration Reform Act abolished country-of-origin quotas that favored European

---

11. Frimpong-Ansah, *Vampire State*, 140.

12. Brydon, "Ghanaian Responses."

13. These SAPs have tended to worsen economic and social conditions in many countries such as Ghana. In 1998 for instance, Ghana spent 75 million USD on social services, which represented 20 percent of the money spent to repay loans (Sassen, "Global Migration"). In this context, alternative survival strategies emerge, such as labor migration to the global North and particularly the United States.

14. Brydon, "Ghanaian Responses," 570. Official estimates of Ghanaians expelled from Nigeria in 1983 are between 900,000 and 1.2 million (ibid.) With the national population of Ghana approximately 10 million in the mid-1980s, this migration from Nigeria brought 10 percent of the population back into the country in the midst of a food and commodity shortage in Ghana.

15. Peil estimated that 20 percent of Ghanaian citizens lived outside of Ghana in 1995 ("Ghanians Abroad," 349), while van Dijk gives an estimate of 12 percent in 1999 ("From Camp to Encompassment"). More conservative estimates were given between 10 percent and 15 percent in 2004 (Wong, "Gendered Politics," 357).

immigrants, thereby increasing opportunities for Ghanaian immigration. Second, the Diversity Visa Program of the 1990 US Immigration Act offered much greater opportunity to Ghanaians.[16] Third, increased restrictions on immigration to Western Europe have led the United States to become the favored destination of Ghanaian immigrants.[17] These policies opened opportunities for Ghanaian immigrants to settle in the United States. Between 2003 and 2006, the number of Ghanaians gaining permanent residency in the United States more than doubled.[18] By 2007, the total number of Ghanaians living in the United States was over 150,000.[19]

The majority of the Ghanaian immigrants in North America are from southern Ghana, are ethnically Akan, and are Christian. From my own estimates, 80 percent are from the Eastern Region (ethnically Akwapim, Kwahu, and Akyem) and Asante: the children, grandchildren, and great-grandchildren of the intrastate cocoa migration in the first half of the twentieth century.[20] Like their cocoa-growing forbearers, they have experienced extreme stress and insecurity as well as shifting matrilineal and conjugal relationships. Many have turned to enchanted Christianity like their cocoa migrant ancestors. Since the turn of the twenty-first century, however, these labor migrants have found relief within the Presbyterian Church, which has branches in several major US and Canadian cities.

---

16. The Diversity Visa Program is an immigration lottery program. Every year the United States gives a certain amount of visas to a specific country. Citizens submit their names to the lottery, and if chosen are given a visa.

17. Olupona and Gemignani, "Introduction," 2. Large numbers of immigrants in Europe by the 1990s, combined with slowing European economies, generated an anti-immigrant backlash. This led to the racialization and criminalization of African immigrants as well as the tightening of European migration policies (Adogame and Weisskoppel, "Introduction," 4).

18. In 2003, the number was 4,410 compared with 9,367 in 2006. Table 3: Persons Obtaining Legal Permanent Resident Status by Region and Country of Birth: Fiscal Years 1997 to 2006, Office of Immigration Statistics, US Homeland Security.

19. This number was calculated from the number of Ghanaian residents in the United States, 104,842 in 2007, and an estimated number of undocumented Ghanaians in the United States, 50,000 for a total of 115,572 (Terrazas, "African Immigrants," 1; Twum-Baah, "Volume and Characteristics," 65).

20. This figure is supported by foreign remittance patterns from abroad, which comprise 40 percent of total remittances, a majority of which come from the United States. The Asante, Eastern, and Greater Accra Regions receive the most remittances (Mazzucato, van den Boom, and Nsowah-Nuamah, "Origin and Destination," 150–51). I speculate that the remittances to Accra are sent primarily to non-Ga migrants in the capital city.

Since the mid-1990s, a network of Ghanaian Presbyterian churches—some affiliated with the Presbyterian Church of Ghana, some with the Presbyterian Church of the USA, and others with the Presbyterian Church of Canada—have become established in North America (see map 4.1).[21] These Ghana congregations of the Presbyterian Church of the USA and the Presbyterian Church of Canada are members of the Conference of Ghanaian Presbyterian Churches, North America. The conference began to organize itself when pastors from the Brooklyn church and the Columbus, Ohio, church, who were friends from Ghana, decided to organize Ghanaian Presbyterian churches in North America. The first official meeting of the conference was in 2000. As of 2011, there were eleven member churches located in Atlanta, the Bronx, Brooklyn, Columbus, Ohio, Houston, Langley, Maryland, Montreal, Philadelphia, Toronto, and Woodbridge, Virginia.[22] There are also three corresponding members—nonmember churches that participate in the conference—that are affiliated with the Presbyterian Church of Ghana. These churches are located in the Bronx, Manhattan, and Worcester, Massachusetts.

Healing and Deliverance in Ghanaian Immigrant Churches in the United States

Not all Ghanaian immigrant churches in the United States are equally enchanted; the Presbyterian churches are particularly devoted to healing and deliverance. In Philadelphia, for example, there are three

---

21. The first Ghanaian Presbyterian community in the Diaspora was established among Ghanaian students in London. By October 1961, the Presbyterian Church of Ghana sent the Rev. F. W. K. Akuffo to London to cater to the spiritual needs of the more than three thousand Ghanaian students residing in London. In November 1961, Akoffo was installed as an assistant minister of the Oxendon Church in London and chaplain to the Ghanaian students in London (Presbyterian Church of Ghana, Reports for 1961, 84). There continue to be Ghanaian Presbyterian churches in greater London, although none have developed prayer teams as of 2007. Author interview with Abboah-Offei, Akropong, February 12, 2007.

22. These include the Bethel Presbyterian Reformed Church (Brooklyn), Christ the King Presbyterian Church (Laurel), Ebenezer Presbyterian Church (Woodbridge), Emmanuel Presbyterian Reformed Church (Bronx), Ghanaian Presbyterian Church (Montreal), Ghanaian Presbyterian Church (Toronto), Presbyterian Church of Ghana Fellowship (Atlanta), Ramseyer Presbyterian Church (Columbus), Trinity Presbyterian Church Fellowship (Bronx), United Ghanaian Community Church (Philadelphia), and the United Ghanaian Presbyterian Church (Houston).

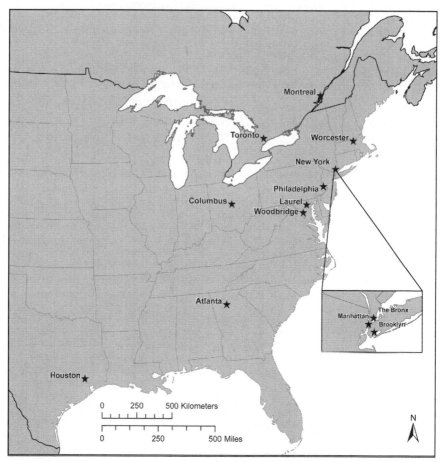

Map 4.1. Location of Ghanaian Presbyterian churches in North America, which includes congregations affiliated with the Presbyterian Church of Canada, the Presbyterian Church of Ghana, and the Presbyterian Church of the USA, 2009. Map by Christine Murray.

Ghanaian churches with over fifty members: the Assemblies of God, Church of Pentecost, and the United Ghanaian Community Church (UGCC), which is Presbyterian.[23] The Assemblies of God is led by a Ghanaian pastor, whereas the congregation is Ghanaian, Liberian, and West Indian. The Church of Pentecost and UGCC have both

---

23. There are also smaller Ghanaian churches such as the Apostolic Church, Seventh Day Adventist, and Lighthouse Chapel International. These data were collected in 2006 and 2007.

Ghanaian leadership and laity. The two Pentecostal churches (Assemblies of God and Church of Pentecost) have an approach to healing and deliverance very different from that of the Presbyterian Church.

While healing and deliverance have become common practices within Ghanaian Christianity, pastors, congregations, and denominations have responded to this phenomenon differently. Most believe that the devil, acting through demons, is ultimately responsible for sickness or misfortune, although sin enables the devil to afflict individuals. Many Ghanaian Pentecostals believe that while a Christian can be attacked by demons, this attack should be combated by individual prayer.[24] Some say that if one is a "true" Christian, one cannot be harmed by the devil. Both the Assemblies of God and the Church of Pentecost in Philadelphia fall into this category.

Pastor Ofosu-Boakye of Philadelphia's Assemblies of God claimed that too much emphasis has been put on deliverance and not enough on the Bible within Ghanaian Christianity. Before he presents his Sunday service to the Assemblies of God congregation, the sick are called forward and prayed for, but there are no deliverance specialists, deliverance workshops, or deliverance services. Pastor Ofosu-Boakye rarely casts out demons, and there is little evidence of demonic possession within his congregation. Pastor Ofosu-Boakye argues that many Christians are looking unnecessarily for demonic explanations to the naturally explained misfortunes and illnesses in their lives.[25]

A similar description came from an elder in Philadelphia's Church of Pentecost. Healing and deliverance are established features of the Church of Pentecost in Ghana. This church emerged during a 1953 Latter Rain revival in Ghana when many demons were cast out, the sick were healed, and accounts were given of the dead being brought back to life. But today there are a variety of opinions concerning this practice within the Church of Pentecost. Many members of the church, including the international director, Rev. Apostle S. K. Baidoo, believe that deliverance is an overemphasized practice in Ghanaian Christianity.[26] This stance of Rev. Baidoo is a result, to an extent, of the resignation in October 1995 of the

---

24. Asamoah-Gyadu, *African Charismatics*, 165–87.

25. Author interview with Pastor Ofosu-Boagye, Assemblies of God, Philadelphia, October 24, 2005.

26. Interview with Elder Godfrey Asiedu, Church of Pentecost, Philadelphia, November 16, 2005.

former leading deliverance practitioner Owusu Tabiri of the Church or Pentecost, who operated the Bethel Prayer Camp in Sunyani, to found the Bethel Prayer Ministries International.[27] It was this division by the church's leading deliverance practitioner, in my opinion, that has contributed to the leadership of the Church of Pentecost to be wary about deliverance practitioners. Like Pastor Ofosu-Boakye, Rev. Baidoo believes that one need only pray to receive healing or deliverance. People of this school within the Church of Pentecost believe that deliverance practitioners are rarely necessary and that no demons can harm people who truly believe in Christ.

There are indications that the Church of Pentecost is beginning to become more supportive of healing and deliverance internationally. For example, In November 2007 Prophetess Boateng from the Church of Pentecost's Edumfa Prayer Camp in Cape Coast conducted healing and deliverance services at several Church of Pentecost branches in North America, including Philadelphia.[28] Her focus was on prayer and repentance to cast out demons, enabled by individual sin, which caused much illness and misfortune to church members in the diaspora.

The leadership of the Ghanaian Presbyterian network in North America is overwhelmingly pro-deliverance. There are two aspects of the healing and deliverance ministry of the Ghanaian Presbyterian churches in North America that set them apart from other Ghanaian churches in the diaspora. One, the Presbyterian churches have a separate group within the church that manages the health and welfare of church members. This group is called the prayer team, which is the BSPG of these Ghanaian Presbyterian churches in North America.[29] Two, these prayer teams are formally trained by Catechist Ebenezer Abboah-Offei. These prayer teams study a deliverance manual written by Abboah-Offei in their home churches, and receive training from Abboah-Offei during his evangelism trips to the various Presbyterian churches in North America. The most influential feature, however, is that the prayer teams participate in an annual deliverance workshop

---

27. Tabiri left the Church of Pentecost because of restrictions placed on his evangelism and fundraising efforts (Gifford, *African Christianity*, 108–9; van Dijk, "From Camp to Encompassment," 147–48).

28. Edumfa was the Church of Pentecost's first prayer camp incorporated into the church as early as the 1960s (van Dijk, "From Camp to Encompassment," 147–48).

29. As in the Ghanaian churches discussed in chapter 3, these groups are alternatively called Prayer Warriors or Prayer Power.

run by Abboah-Offei in New York.[30] By the early to mid-2000s, these Ghanaian Presbyterian churches began developing robust healing practices in North America as an extension of such practices in the Presbyterian Church of Ghana.

Transnational Deliverance Ministry and Workshop, from Ghana to North America

A network of kinship ties facilitated Abboah-Offei's transnational deliverance ministry, as it did locally in Ghana. Abboah-Offei's wife, Faustina Abboah-Offei, is the niece of Paulina Atiemo, who is married to the Rev. Samuel Atiemo, Pastor of the Bethel Presbyterian Reformed Church in Brooklyn. Both Abboah-Offei and Atiemo come from Akropong Presbyterian families, and their fathers and grandfathers were cocoa growers in Akyem Abwakwa. Both spent parts of their childhoods in Asamankese, hearings stories about *kyiribentoa* (the Twi name for Faith Tabernacle), the church that abjured biomedicine for healing. Both men married into the same *abusua* (matrilineage). While not childhood friends, they crossed paths before their marriages through their involvement in various parachurch organizations in Asamankese.

Abboah-Offei knew Atiemo from Asamankese, where Atiemo was born and Abboah-Offei went to secondary school. By the early 1970s, Atiemo was involved in a number of parachurch organizations including Scripture Union and Joyful Way. Samuel Atiemo was the first external missionary outreach program director of Joyful Way by 1978.[31] Catechist Abboah-Offei was the Joyful Way prayer coordinator in Asamankese, and through this organization the two met. Abboah-Offei recalls that Atiemo was like a big brother to him, later financing and organizing his wedding. After Abboah-Offei's marriage to Faustina, he and Atiemo got to know each other much better, as their children became part of the same matrilineage.

---

30. At the deliverance workshop in 2007, the chairman of the Conference of Ghanaian Presbyterian Churches, North America, Dr. Rev. Ofosu-Donkoh, argued that the prayer teams within the Ghanaian Presbyterian churches and their training during the deliverance workshop were both essential to fulfill the evangelism efforts of the Conference of Ghanaian Presbyterian Churches, North America.

31. Barker and Boadi-Siaw, *Changed by the Word*, 166.

In October 1993, Atiemo moved to Brooklyn after attending Fuller Theological Seminary in Pasadena, California, and helped to establish the Bethel Presbyterian Reformed Church.[32] A year later, Atiemo visited his in-laws in London, where he discussed the slow growth of his Brooklyn church with Joseph Martinson, Paulina Atiemo's nephew and Faustina Abboah-Offei's brother. Martinson suggested that Atiemo contact Catechist Abboah-Offei and invite him to host a revival in order to boost the membership of his Brooklyn church. This was the first time Atiemo had heard of Abboah-Offei's deliverance ministry. A few years later, Atiemo contacted Abboah-Offei about coming to Brooklyn and holding a healing and deliverance revival. Unfortunately, Atiemo was not able to finance the trip.

Fortuitously, in 2001, the chairman of the Akropong Presbytery wrote a letter to Atiemo explaining that he wanted Catechist Abboah-Offei to solicit funds for the Presbyterian University in Akropong through a series of healing services within the Ghanaian Presbyterian Churches in North America. This fundraising project took a few years to develop, but eventually Abboah-Offei travelled to the United States in the summer of 2004 for eight weeks to perform deliverance services and individual consultations for suffering Ghanaians, as well as to solicit funds for the Presbyterian University. During this trip, Abboah-Offei visited Ghanaian Presbyterian congregations in the Bronx, Brooklyn, Columbus, Ohio, Houston, Manhattan, Woodbridge, Virginia, and Worcester, Massachusetts. He also visited Ghanaian Neo-Pentecostal churches in Maryland and Chicago. Not only did the membership of the host churches increase, but numerous people were also healed of various maladies.

Abboah-Offei finished his 2004 evangelism trip with a brief stay at the Presbyterian retreat center in Stony Point, New York, from August 29 through September 3. During this informal gathering, Abboah-Offei taught some prayer-team members from the New York congregations, particularly Samuel Atiemo's church in Brooklyn, and ministered to the afflicted through individual consultations. This five-day retreat was so successful that Abboah-Offei and Samuel Atiemo decided to host an annual deliverance workshop at the Stony Point Retreat Center to train all the prayer teams from the Ghanaian Presbyterian Churches in North America.

---

32. Before attending Fuller, Atiemo had been the Africa area director of Youth for Christ International (a parachurch organization established by Billy Graham) from 1982 to 1992.

After this US evangelism trip in 2004, many of the US Ghanaian Presbyterian churches invited Abboah-Offei for return visits, and many non-Presbyterian congregations also invited him to conduct services. He traveled to the United States and Canada in 2005 for ten weeks, again to perform deliverance services and raise money for the Presbyterian University.[33] During the 2005 evangelizing trip, the catechist, in addition to revisiting the congregations he had seen in his 2004 trip, visited other Ghanaian Presbyterian churches in Montreal, Philadelphia, and Toronto, as well as Ghanaian Neo-Pentecostal churches in Newark and Washington, DC. In November 2005, Abboah-Offei offered a deliverance workshop for Ghanaian Presbyterian Churches (Presbyterian Church of Ghana, Presbyterian Church of Canada, and Presbyterian Church of the USA) in Stony Point, New York.

The primary goal of the workshop was to teach individual prayer teams at each Ghanaian Presbyterian Church in North America how to minister to the afflicted within their own congregations.[34] This deliverance workshop in New York was the first attempt to fully replicate the deliverance course developed by Abboah-Offei and taught at Grace Presbyterian Church. As in Ghana, the catechist was concerned with people consulting non-Christian healing practitioners or leaving the Presbyterian Church for a more enchanted church. Abboah-Offei was particularly troubled by the large number of Ghanaians who served various deities in order to gain protection before immigration and who brought various protective charms (*suman*) with them when migrating.[35] Therefore, he wanted to establish healing and deliverance practices firmly within the Ghanaian Presbyterian Churches in North America.[36]

---

33. During a similar fundraising trip in 2005, Abboah-Offei raised $3,500 from the Ghanaian Presbyterian Church in Philadelphia (UGCC), $2,250 from Atiemo's Brooklyn congregation, and $2,150 from the Ghanaian Presbyterian Church in Columbus, Ohio ("3 Ghanaian PCUSA Churches Donate To Presby University, Ghana." Ghanaweb, Diasporian News, Friday, July 14, 2006).

34. Abboah-Offei did not create the individual prayer teams, however. This was done locally at each church in coordination with their pastor.

35. Abboah-Offei said that many Ghanaians, when returning to Ghana, pay tribute to particular shrines. Much to his family's displeasure, one of Abboah-Offei's cousins has established a shrine to a deity in New York. Author interview with Abboah-Offei, Stony Point, November 26, 2007.

36. Data from Grace Presbyterian Church show that within a case study of fifteen thousand Christians who have undergone consultation, 75 percent had previously consulted some other form of religious healing practitioner outside of the church.

With this in mind, Rev. Atiemo was able to reserve space for the meeting at the Presbyterian Retreat Center in Stony Point, New York. The timing of the New York workshop in late November allowed Abboah-Offei to return to Ghana to teach the Scripture Union deliverance course during the last week of December in Aburi and Kumasi, as well as his own deliverance course in Akropong, which began in mid-February. In 2005, prayer teams members from Brooklyn, Bronx, Columbus, Houston, Manhattan, Montreal, Philadelphia, Toronto, Woodbridge, and Worcester attended, as well as various others who sought deliverance through personal consultation with the Catechist.

Most of these suffering Ghanaians (non-prayer-team members) came from greater New York, where buses connect Manhattan to Stony Point, although others came from much farther away.[37] One person who attended the workshop in 2005 flew from Houston for one night. Of the healings that occurred in 2005, the most memorable was for a woman with cancer for whom everyone prayed. After she returned home from the workshop, her doctors could no longer detect the disease.[38] The large number of suffering people who came for deliverance required so much of Abboah-Offei's time that very little deliverance instruction was given to the prayer teams. Abboah-Offei and some of the other participating pastors decided to reduce the numbers of counselees by restricting the participation to only five prayer-team members from each church in 2006. Rev. Atiemo was the chief facilitator of the Stony Point deliverance workshop in 2006, as well as the 2006 president of the Conference of Ghanaian Presbyterian Churches, North America.

Course Setting, Course Structure, and Course Participants

The Stony Point Center, one of three national conference centers of the Presbyterian Church USA, is located on 32 acres of land in the Lower Hudson River Valley between New York City and West Point. Originally used as a missionary training center, in 1977 the Stony

---

This figure was given publicly during a deliverance service at the United Ghanaian Community Church in Philadelphia by Abboah-Offei on April 25, 2006.

37. During Abboah-Offei's week-long deliverance program in Philadelphia in September 2005, Ghanaians came from as far as Florida to have Abboah-Offei consult, pray, and deliver them from various illnesses.

38. Interview with the catechist Ebenezer Abboah-Offei, Akropong, February 15, 2007.

Point Center for Education and Mission was established by the Presbyterian Church USA as a retreat and conference center for Christians from all over the world.[39] Stony Point aimed to become a global Christian village. This aim was achieved when the Ghanaian Presbyterian Churches in North America began meeting there annually in 2005 for a deliverance workshop led by Catechist Abboah-Offei and his assistant Samuel Asare.

The 2006 workshop took place from Sunday, November 26, to Saturday, December 2. The first Sunday was reserved for arrival and registration. The schedule from Monday through Friday followed the same pattern. Many of the prayer teams woke at 3 am and prayed in groups until 6 am, typically in one of the dorm rooms. From 6 am to 8 am, all the participants gathered for praise and worship, which consisted of individual praying and group prayer, as well as singing and dancing. Group devotion occurred from 8 am to 9 am, when the participants would pray together about certain topics designated by a prayer leader. From 9 am to 12 pm, various deliverance lectures/sermons were given by Samuel Asare and conference participants. At 12 pm there was a half-hour break, followed by another three-hour block of lectures/sermons. From 3:30 pm to 6 pm there was group prayer, and from 6 pm to 7 pm, dinner, which was the only meal of the day taken by participants. The period from 7 pm to 11 pm was allotted to lectures/sermons, as well as individual consultation with Samuel Asare for participating members. This was the basic structure of the workshop from Monday through Friday, and on Saturday, December 2, the prayer teams left Stony Point to return to their home cities.

At the deliverance workshop in 2006, at least one member was present from each of the prayer teams representing Brooklyn, the Bronx,[40] Columbus, Houston, Manhattan, Philadelphia, Toronto, Woodbridge, and Worcester. In attendance were also several non-prayer-team members: at least one visitor from Ghana, an anthropologist (the author), and a few people who came for individual consultation. Catechist Abboah-Offei was unable to attend the 2006 workshop owing to exhaustion from having held deliverance services

---

39. Flory, "Dear House," 40.

40. There are two Ghanaian Presbyterian Churches in the Bronx, one affiliated with the Presbyterian Church USA and called the Presbyterian Church of Ghana Mission, and one affiliated with Presbyterian Church of Ghana and called Presbyterian Church of Ghana. Both churches had prayer-team members in attendance.

in Toronto and Montreal during the prior two weeks. In his place were his assistant, Samuel Asare, who works with the catechist at the Grace Presbyterian Church, and the host, Rev. Samuel Atiemo. There were approximately eighty to one hundred adults attending the course throughout the week, although typically only half that number was present at any given time. More than 60 percent of the participants were woman.[41] The ages of the participants ranged from twenty-three to sixty-six and were fairly evenly distributed across this range.[42] Almost half of the group had immigrated to North America in the 1990s, and over 70 percent had immigrated in the 1990s or 2000s.[43] Of the workshop participants, 50 percent worked in the healthcare industry, and over 70 percent had some postsecondary education either in Ghana or North America.[44]

In terms of religious experience and affiliation, 70 percent of the participants had never been to Grace Presbyterian Church or any other Christian deliverance center or prayer camp in Ghana.[45] More than 50 percent of participants had been involved in Scripture Union in Ghana as members, presidents, camp leaders, speakers, or praise and worship leaders in both secondary school and in town fellowships.[46] Over 70 percent of respondents had been members at one time of the Presbyterian Church of Ghana before migrating to North America.[47] Over 75 percent of participants joined their local

---

41. Thirty of forty-nine participants were woman.

42. Eight participants were in their twenties, twelve participants were in their thirties, seven participants were in their forties, ten participants were in their fifties, and six participants were in their sixties.

43. Eight immigrated in the 1970s, six in the 1980s, twenty-three in the 1990s, and twelve in the 2000s.

44. Twenty-five of fifty respondents worked in some capacity in healthcare. Responses included a drug abuse therapist, mental retardation worker, nurses assistant (CNA), nurses aid, healthcare provider at a nursing home, home health assistant, home health aid for the elderly, nurse (LPN), pharmacy technician, extended care provider, physical therapy assistant, mental health care provider, and preoperative clinic coordinator. Thirty-five of forty-nine respondents had some form of postsecondary education.

45. Thirty-five of fifty had never been to Grace or any other healing center, while nine of fifty had visited Grace.

46. Twenty-six of fifty-one were affiliated in some capacity with Scripture Union in Ghana.

47. Twenty-four of fifty-one had only been members of the Presbyterian Church of Ghana, while thirteen of fifty-one had been members of the Presbyterian Church of Ghana as well as at least one other church. Other churches listed included Church of

Figure 4.1. Participants in the New York deliverance workshop on November 26, 2007. Catechist Ebenezer Abboah-Offei is holding a notebook and walking down the center aisle. Photograph by the author.

Ghanaian Presbyterian Church upon migrating to North America.[48] Some of the others joined native-born congregations such as Baptist, Catholic, or Methodist churches, while three had been members of Lighthouse Chapel in North America, a branch of the Accra-based Neo-Pentecostal Church.

Deliverance literature also played a role in the workshop. On the last page of Abboah-Offei's deliverance manual is a list of deliverance books, which he recommends for extra reading to supplement the deliverance practitioner's knowledge. The authors listed include Rebecca Brown, Mark Bubeck, Abraham Chigbundu, Peter

---

Pentecost, Seventh Day Adventist, Catholic, Methodist, Anglican, Apostolic Church, Musama Disco, United Pentecostal Church, Assemblies of God, and Baptist. Several Charismatic Pentecostal churches were listed such as GLWC, Lagos-Rhema, Good News Bible Church, House of Faith, and Mount Calvary Cross Ministries.

48. Thirty-nine of fifty-one respondents said they joined their local Ghanaian Presbyterian church upon migrating to North America.

Horrobin, Harold Horton, Francis MacNutt, Ed Murphy, Samuel O. Onwona, and Dean Sherman.[49] Of the deliverance workshop participants, 30 percent were frequent readers of deliverance literature.[50] While the authors listed by Abboah-Offei wrote the texts primarily read by participants, books by the early Pentecostal healing evangelist Smith Wigglesworth (1859–1947) became popular among the participants during the course of the workshop.

Two of the workshop participants referred to Smith Wigglesworth's ministry during morning lectures given by participants. One of these participants, John Gyadu (pseudonym), particularly discussed Wigglesworth's faith; Wigglesworth was an evangelist who believed that relying on biomedicine was unscriptural and that having faith in God alone was sufficient for healing.[51] Gyadu testified to abstaining from all treatment but prayer for the past two years. After Gyadu's testimony about the power and importance of Wigglesworth's ministry, which could be found in a variety of books written by and about him, many of the other participants began downloading Wigglesworth's writings from the Internet and ordering his books online.

Confrontation and Deliverance during the New York Workshop

The deliverance workshop had both theoretical and practical components in 2006, 2007, and 2008. While theories of disease, misfortune causation, spiritual attacks, and demonic possession were discussed during the lectures, which follow a booklet written by Abboah-Offei, there was a practical component as well. The practical component consisted of teaching participants how to become deliverance counselors. This individual consultation procedure was taught to participants in two ways: through experiencing the process as a counselee and by participating as a counselor with one or more trained deliverance practitioners.

---

49. Abboah-Offei, *Church Leaders Training Manual*, 106.

50. Fifteen of fifty respondents said they frequently read deliverance literature and listed their favorite authors. Five of fifty respondents listed Rebecca Brown as a particularly influential deliverance author. Other authors listed more than once were Francis MacNutt, Frank and Ida Mae, and Benny Hinn.

51. Before converting to Pentecostalism, Wigglesworth was part of the larger divine healing movement that included Faith Tabernacle, which abstained from using medicine.

Everyone participated as both counselors and counselees during the workshop in both years. All participants attending the workshop filled out the deliverance questionnaire. Every evening several people met with Ebenezer Abboah-Offei, Samuel Asare, and Samuel Atiemo to participate in counseling sessions as counselors and counselees during the allotted time for "clinic," as the counselor-training segment was called. The counseling sessions occurred in private rooms off the main conference center. Typically this form of consultation takes place privately with only the counselor(s)—various members of the prayer teams—and the counselee present. These consultations were semipublic during the workshop in order to train participants to become counselors.

Deliverance practices were also performed during lectures and sermons in response to spiritual attacks on conference participants. On two separate occasions in 2006, demonic spirits attacked workshop participants: one with epileptic seizures and another through a near accident. Subverting the afflicting agents required the prayer efforts of all the workshop participants.

On the morning of November 28, 2006, Samuel Asare gave the morning lecture at Stony Point. Asare announced that during the previous night one of the participants had been attacked by witches, and he gave some background to the case. Kofi Acheampong (pseudonym) had been attending prayer meetings at his home church to receive healing for his epilepsy. The prayer team at his home church in Houston suggested that Acheampong come to Stony Point and be seen by Abboah-Offei for relief from his condition. For reasons unknown, Acheampong left his medication in Texas, which greatly angered Asare, who, like Abboah-Offei, constantly preaches the incorporation of biomedicine.

During the first night at Stony Point (November 27) Acheampong had mild seizures. The prayer team from his church prayed vigorously for him to recover. When the seizures ceased, the Houston prayer-team members thought he had recovered. Sometime in the middle of the second night, however, Acheampong had more serious seizures, began to bleed, and was rushed to the hospital. Many at the workshop concluded that his seizures were caused by an agent located in the spiritual realm. Acheampong was bewitched.

After explaining this course of events the following morning to the entire workshop, Asare led the sixty-plus participants in a prayer line, praying against witches, the diagnosed agent responsible for

Acheampong's seizures. The series of prayers constituting the witchcraft prayer line are found within Abboah-Offei's deliverance manual. An example from the witchcraft prayer line (6) is as follows: "break all witchcraft alliances, links and communication networks and bum [sic] all gadgets with the fire of God."[52] Once Asare read the prayer, everyone prayed about this issue individually. All participants prayed out loud together with varying tones and pitches. Each of these prayer topics would last for five to ten minutes. When the volume of prayers began to decrease, Asare read the next prayer line. This prayer line against witchcraft attacks lasted almost two hours.[53]

The following night another spiritual attack occurred. During the night of November 28, a pregnant women named Mary Anim (pseudonym) fell asleep in her room. She had put a towel over a lamp in order to create a dimmed light in the room. Early in the morning she awoke to the fire alarm, and immediately noticed that the towel was smoking. Anim rushed the lamp and towel outside. Once outside the towel immediately caught fire. On the morning of November 29, Samuel Asare interpreted the events that unfolded the previous night.

Unlike the morning before, Asare did not interpret this event as a direct witchcraft attack. The event was interpreted as the replication of a specific biblical demonic attack recounted in Revelations 12:1–4. Asare reported: "If it had not been for God, a baby and its mother would have been swallowed by a great red dragon." This biblical passage tells of a pregnant woman confronted by a red dragon, which meant to consume the pregnant woman's child. This passage was drawn on a large sheet of paper for the course participants (see fig. 4.2). The biblical red dragon that attacked the pregnant woman the previous night, argued Asare, was the devil and acted as the accomplice of witches.[54] Witches are believed to feed on children and unborn babies, particularly within one's own family.[55] This interpretation by Asare was followed by a deliverance lesson about

---

52. Abboah-Offei, *Church Leaders Training Manual*, 105.

53. It is important that Asare interrupted the prayer line once to make sure that everyone understood that they were praying against supernatural witches and not real people, such as aunties or grandmothers.

54. It is common for several spiritual agents to be involved with someone's illness or misfortune within this deliverance system.

55. In Twi, the etymology of witch (ɔbayifo) is derived from *oba*, meaning child, and *yi*, to take away (Christaller, *Dictionary*, 11). Therefore, witchcraft in Ghana particularly, but not exclusively, affects pregnant women and children.

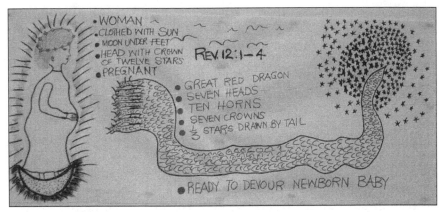

Figure 4.2. Drawing by Samuel Asare at the New York deliverance workshop on November 29, 2006, of the biblical red dragon ready to consume a woman's unborn child. Photograph by the author.

shedding demonic bondage and was finished with a prayer line against demonic altars. Demonic altars are structures on which sacrifices are made, curses are enacted, and demons or witches consume victims.[56] Destroying the power emanating from the altar—considered the source of the afflicting power—through prayer was meant to free Anim from the danger posed to her and her baby by the devil and by witches.

Samuel Asare indicated on the morning after Mary Anim's attack that the targets of the devil or witches were not only the woman and child, who were nearly consumed by fire, but also all participants in the course. One message consistently reiterated during the lectures was that participation in the deliverance ministry abounds with demonic confrontation. Samuel Asare stated, "You will not achieve your goals on a silver platter. You have to struggle. Whenever there is something good coming, you will be confronted by the evil one." These statements were most visibly realized by the spiritual attacks on Kofi Acheampong and Mary Anim. But during the course of the lectures, as well as the praise and worship sessions, there were other signs of demonic possession in the form of demonic manifestations. Abboah-Offei writes that manifestations are signs of the presence of a spirit or demon; they take various forms during the process of deliverance. There are sixty-two signs listed, some of which include

---

56. Abboah-Offei, *Church Leaders Training Manual*, 37–38.

climbing on objects, removing of clothes, hitting one's head on the floor, screaming, fainting, hissing, snarling, or spitting.[57]

During the course of preaching, praying, singing, or dancing, many people were prayed for or had hands laid on them. If possessed, demons resist expulsion and therefore cause various kinds of involuntary movements or elicit utterances from their host (see fig. 4.3). Some people's eyes roll back in their heads. Others fall over. Some scream and jump. Others get violent, attacking the people praying for/with them. Many collapse to the ground and often vomit different types of objects and substances. This often-violent process of expelling demons from Christian bodies and replacing them with the Holy Spirit was consistently being performed throughout the week of the deliverance workshop.

Deliverance Replications in the Diaspora

A number of beliefs and practices with respect to healing and deliverance have been replicated through the New York deliverance workshop.[58] Prayer teams at individual churches in the United States and Canada have been trained through a course taught at the Grace Presbyterian Center by Abboah-Offei. Through this course, the BSPG of the Presbyterian Church of Ghana has been replicated in the prayer teams in the diaspora. The individual counseling session has also been maintained transnationally, used not only during the deliverance workshop but also by the prayer teams of the various Ghanaian Presbyterian churches. The methods of laying on hands, anointing, and prayer are the same in Ghana and the United States. When a specific attack occurs, a diagnosis is determined, and intercessors gather and read specific prayer lines against the specific afflicting agent. All these institutions and practices have been replicated in the Ghanaian Presbyterian Churches in North America through the New York deliverance workshop.

The primary diagnostic tool, the deliverance questionnaire, is used in the diaspora. The afflicting agents are the same both in both Ghana and the United States, as demonstrated by the cases of Kofi

---

57. Ibid., 27–29.

58. These observations are based on fieldwork in Philadelphia from July 2005 through July 2009, the Stony Point Deliverance Workshop in November 2006, 2007, and 2008 at Grace Presbyterian Center in February 2007.

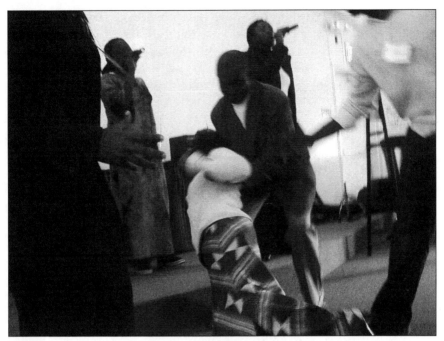

Figure 4.3. Woman possessed at New York deliverance workshop on November 27, 2006. Photograph by the author.

Acheampong and Mary Anim. Witches continue to cause illness and misfortune in the Ghanaian Presbyterian community in North America. As in the Presbyterian community in Ghana, witchcraft attacks on pregnant women are common.[59] Witches are interpreted in the diaspora as the devil or the accomplices of the devil, as they are in Ghana.[60] As Samuel Asare discussed in one of his New York lectures, the majority of afflictions listed on the questionnaires he examined in the diaspora were obvious witchcraft attacks.[61] The afflicting agents, such as witches, are almost always located in Ghana, not abroad.[62] This too is carried over from deliverance practices in Ghana, where

---

59. Witchcraft attacks on pregnant women are very common in Ghana. On February 9, 2007, during a deliverance service at Grace Presbyterian, Abboah-Offei called all pregnant women forward to lay hands on their unborn babies and pray against witchcraft attacks.

60. See Meyer, "'If You Are a Devil.'"

61. Samuel Asare lecture, Stony Point, New York, November 28, 2006.

62. The one exception was an accusation against Haitians and Puerto Ricans for using "black magic"—probably referring to Vodun and Santeria—in order to beat

people are not typically afflicted by agents outside of Ghana. And witchcraft attacks among these labor migrants, as during the colonial period in the midst of the cocoa migration (see chapter 2), most typically emanate from within the sufferers' own families and are related to issues of resource distribution.[63] Major disjunctures in the transnational practice of deliverance, with particular respect to gender, is the subject of chapter 6.

Conclusion: Enchanted Calvinism in North America

From the 1960s through the turn of the twenty-first century, a failing economy and unstable government in Ghana led to a large-scale immigration to North America, where many migrants established and joined Ghanaian Presbyterian churches. During this process of migration, spiritual entities—particularly witches—continued to afflict these Ghanaian Presbyterians in North America. Unlike during the cocoa migration in the early twentieth century, the Presbyterian Church of Ghana had a plethora of resources at its disposal in order to combat these evil forces; the church was enchanted. Above all, the charismatic catechist Ebenezer Abboah-Offei was able to address the spiritual needs of the diasporic congregations. Abboah-Offei's transnational healing ministry is not an isolated phenomenon, but rather part of a much larger trend within transnational forms of African therapeutics whereby religious healers, biomedical professionals, and pharmacopeia are in constant circulation between Africa and the outside world,[64] a process that John Janzen describes as "Afri-global medicine."[65]

Specifically, Abboah-Offei facilitated the transplantation of healing and deliverance practices, as well as the institutional framework of such practices (the BSPG), to North America through the creation of prayer teams within the North American Ghanaian Presbyterian Churches. Prayer teams at these various North American churches have been trained by Abboah-Offei in deliverance practices primarily

---

out Ghanaian Christians for jobs. This discussion between Abboah-Offei and several prayer-team members occurred in Philadelphia in April 2006.

63. Ter Haar found a similar pattern among Ghanaian Christian immigrants in the Netherlands ("Ghanaian Witchcraft Belief," 93–112).
    64. Langwick, Dilger, and Kane, *Medicine, Mobility, and Power*.
    65. Janzen, "Afri-Global Medicine."

through a weeklong deliverance workshop held annually in New York since 2005, although some training does take place within individual congregations (see chapter 5). Many aspects of the deliverance ministry have been replicated in the diaspora through the New York deliverance workshop, including diagnostic techniques, individual consultations, and the various ritualized prayers and techniques used to treat afflicted Christians. Today, the Ghanaian Presbyterian Churches in North America are particularly enchanted.

But how are these malevolent agents managed within local Ghanaian Presbyterian congregations in North America? In the following chapter I will narrow my focus to the United Ghanaian Community Church, the Ghanaian Presbyterian Church based in Philadelphia. I will describe the formation of this church, the establishment of the Philadelphia prayer team, and how deliverance is practiced at the congregational level.

5

The Enchantment of the United Ghanaian Community Church, Philadelphia

The primary mechanism for establishing deliverance practices among Ghanaian Presbyterians in North America has been the New York deliverance workshop, held annually since 2004 by Abboah-Offei, with the goal of standardizing these practices among the various Ghanaian Presbyterian congregations. As I demonstrated in the previous chapter, this goal has been generally successful, particularly since the Ghanaian Presbyterians in North America combat the same malevolent forces.

One area of difference within the North America Ghanaian Presbyterian community, however, is the development of prayer teams across the member congregations of the Conference of Ghanaian Presbyterian Churches, North America. By 2008, the Brooklyn church had not installed a formal prayer team, but invited all members to participate during prayer services on Wednesday and Thursday nights. In Columbus, Ohio, a group of six people frequently gather to pray, although a formal and organized prayer team had not yet been formed by 2010. Alternatively, in Philadelphia, a formal and structured prayer team has developed as a result of the evangelizing and training efforts of Abboah-Offei. The development of the prayer team of the Ghanaian Presbyterian Church in Philadelphia—called the United Ghanaian Community Church (UGCC)—as well as the healing and deliverance practices performed within its congregation, is the subject of this chapter. The UGCC is a particularly enchanted congregation within the Conference of Ghanaian Presbyterian Churches, North America.

In this chapter I describe how the UGCC became enchanted and show how spiritual explanations of the world—particularly as it relates to health and healing—became prevalent among Ghanaian labor migrants in Philadelphia, how the lives of these Presbyterians were inhabited by malevolent spirits, and finally, how healing practitioners (i.e., the Philadelphia prayer team) developed to manage the complex array of human-spirit interactions.

The Establishment of UGCC in Philadelphia

The story of the UGCC's founding is largely the story of its founder, Dr. Rev. Kobina Ofosu-Donkoh. Ofosu-Donkoh's experiences in Ghana and in the United States shaped the kind of congregation he would establish. His personal narrative has much in common with Abboah-Offei as well as many others involved in Pentecostalism, Neo-Pentecostalism, and the Charismatic Movement among mainline churches in Ghana: cocoa farming, witchcraft attacks, Faith Tabernacle, Scripture Union, the BSPG, and labor migration.

Kobina Ofosu-Donkoh in Ghana

Kobina Ofosu-Donkoh was born in 1956 in the small cocoa farming town of Breman Asikuma in Ghana's Central Region north of Winneba, where his mother's brother (*wɔfa*) was a local cocoa buyer for the government. Cocoa was the primary industry within his family, his village, and in neighboring villages when Ofosu-Donkoh was a child. While the Central Region was heavily Methodist, Ofosu-Donkoh's father, the family elder (*abusua panyin*), had facilitated the establishment of a Presbyterian branch in Breman Asikuma after attending the Presbyterian boarding school in Nsaba, the oldest Presbyterian church (est. 1891) and school in the Central Region.

But seven days after Ofosu-Donkoh's birth, his mother died, which was interpreted by Ofosu-Donkoh's father to be the result of a witchcraft attack by her jealous sisters. Another reason for this interpretation was that she had five successive miscarriages before Ofosu-Donkoh was born.[1] After Ofosu-Donkoh's birth and his mother's death, his father was forcibly removed, or destooled, as family elder. In response, Ofosu-Donkoh's father withdrew his membership from the Presbyterian Church and, for a while, joined Faith Tabernacle.

---

1. Author interview with Ofosu-Donkoh, Philadelphia, July 26, 2005.

After his mother's death, Ofosu-Donkoh stayed in Breman Asikuma and lived with his father's family, while being cared for by his father's second wife. In order to protect Ofosu-Donkoh from his jealous aunts, Ofosu-Donkoh's paternal family told his maternal family that he had died with his mother at childbirth. While growing up, his paternal uncles never let Ofosu-Donkoh visit his maternal aunts, and in particular to eat with his matrilineage, for fear that he would be harmed or poisoned.[2]

Ofosu-Donkoh, like his brothers and sisters, was baptized and educated in the Presbyterian Church, going to both primary and middle school in Bremen Asikuma. After graduating from high school in 1977, Ofosu-Donkoh left his village for Accra. In Accra, he stayed in the house of his paternal uncle, who was president of Trinity Seminary at the University of Ghana and a very prominent Methodist pastor. Ofosu-Donkoh found employment at the Ghana Customs and Excise Department, and received eighteen months of training before beginning work as a junior customs officer at the Kotoka International Airport. This job was extremely lucrative, as a significant amount of money was earned through bribes. As a result, Ofosu-Donkoh felt guilty and unhappy for accepting money illegally.

In Accra from 1979 to 1984, Ofosu-Donkoh attended Trinity United Church, located on the campus of Trinity Theological Seminary near the University of Ghana. Trinity was an interdenominational church established for university workers. At Trinity United Church, Ofosu-Donkoh became particularly inspired by an Akropong woman named Doris Otiwa Acheampong, who frequently spoke at Scripture Union meetings at Trinity and later socially adopted him. Acheampong was a Presbyterian, and Ofosu-Donkoh remembers her as a very pious and spiritual woman. Through Acheampong, Ofosu-Donkoh became involved in Scripture Union.

While Ofosu-Donkoh was nurtured by Acheampong and became born-again, he felt even unhappier about the immoral aspects of his

---

2. As Maia Green argued for contemporary Tanzania, witches not only eat their victims, but also frequently poison them by putting harmful medicines in their victim's food ("Medicine and the Embodiment," 125). A similar set of circumstances occurred in the early childhood of one of the UGCC's prominent members, Hector Smith, whose wife Margaret is on the prayer team. Smith is from Fanteland but grew up in Kumasi because of the fear that he might be harmed by witchcraft if he stayed in his home village. Author interview with Ofosu-Donkoh, Philadelphia, August 23, 2006.

job. Eventually he decided to fast and pray about his situation for seven days. On the seventh day he went to the sanctuary at Trinity, to be closer to God. During prayer, Ofosu-Donkoh had a vision of two huge hands holding a large open Bible, which was handed over to him. Ofosu-Donkoh shared this vision with the Scripture Union leadership, who interpreted this event as a sign from God that he had been called to preach. Shortly thereafter, Ofosu-Donkoh quit his customs job. Many of his family members accused him of being crazy for quitting such a lucrative position. One family member even suggested institutionalizing him in a psychiatric facility, but Ofosu-Donkoh had decided to become a pastor.[3] The leadership at Trinity United Church facilitated his enrolment in Trinity Theological Seminary at the University of Ghana, the primary Protestant seminary in Ghana.

Ofosu-Donkoh studied at Trinity for three years and in June 1984 was commissioned in the Presbyterian Church of Ghana. He spent the next two years as a junior minister working in the Akyem-Oda district. In June 1986, he was ordained as a full-fledged minister in the Presbyterian Church of Ghana, working as the direct pastor of the Akyem-Asuom district until August 1989. During this time he became involved in the BSPG. Feeling that he needed to further his education, Ofosu-Donkoh began applying to seminaries in the United States.

Ofosu-Donkoh in Philadelphia and the Organization of the UGCC

While in Ghana, Ofosu-Donkoh corresponded with Oral Roberts, the famed healing evangelist of the Neo-Pentecostal movement in the United States. The two men developed a friendship and later Roberts awarded Ofosu-Donkoh a scholarship to study theology at Oral Roberts University in Tulsa, Oklahoma. The scholarship covered tuition, but not room and board. At the same time Pittsburg Theological Seminary offered Ofosu-Donkoh a scholarship, and because it covered tuition as well as room and board, Ofosu-Donkoh moved

---

3. In 1979, after Flt. Lt. Jerry Rawlins led a military coup to overthrow the Ghanaian government, all government officials were replaced. Ofosu-Donkoh would have lost his customs agent job. In the summer of 2005 Ofosu-Donkoh returned to Ghana and met a former colleague at the customs department, who had become very poor. Ofosu-Donkoh believed the coup and its repercussions vindicated his turning to the ministry. Sermon delivered by Ofosu-Donkoh at the UGCC, August 6, 2006.

to Pittsburgh in September 1989 to enroll in the seminary. By 1991, Ofosu-Donkoh had earned a master's degree in church history, theology, and ethics. After graduation, he moved to Philadelphia and enrolled in Temple University's master's program in world religions, and later Temple's doctoral program in comparative religion, which he completed in 2003.

When Ofosu-Donkoh moved to Philadelphia in 1991, he befriended another Ghanaian named Emmanuel Opoku. Opoku introduced Ofosu-Donkoh to several Ghanaians living in greater Philadelphia. He also told Ofosu-Donkoh about two Ghanaian associations in the city: the African International Association, whose membership was mostly Ghanaian, and the Ghanaian Association of the Delaware Valley. These were mutual aid associations that helped Ghanaians settle in the United States, advised them on immigration issues, helped with funeral arrangements, and performed various Ghanaian rites of passage such as outdooring and naming ceremonies for recently born children as well as many other social functions.[4] Most of the Ghanaians Ofosu-Donkoh met did not go to church.[5] In the early 1990s there were no Ghanaian immigrant churches in Philadelphia, although there were two Nigerian immigrant churches.[6]

Ofosu-Donkoh decided to gather more information by distributing a survey to Ghanaians at various social gatherings in the summer of

---

4. Much of the power and influence of ethnic associations and local chieftaincies has been usurped by the African immigrant churches, particularly with regard to performing rites of passage such as outdoorings and naming ceremonies, weddings, and funerals. Ofosu-Donkoh recalled performing a naming ceremony in the UGCC in 2005. A local chief was told he could not pour libation in the church and subsequently refused to come to the ceremony. Author interview with Ofosu-Donkoh, Philadelphia, September 9, 2005.

5. Before the UGCC was established, some of the Ghanaians in Philadelphia worshipped in American churches led by Ghanaian pastors. Rev. Dr. Andoh pastored St. Marys/St. Andrews Episcopal Church in West Philadelphia on 36th and Barrow St., and was previously an Anglican pastor in Ghana before earning a PhD at Princeton Theological Seminary. Another Ghanaian pastor, Rev. Kyeremantang, led a Church of Brethren congregation in Philadelphia after attending Lancaster Theological Seminary. Both Ghanaian pastors had several Ghanaians in their congregations. After the UGCC was established in 1996, most of these Ghanaian members joined the UGCC.

6. The first two African immigrant churches in Philadelphia were Nigerian: the Celestial Church of Christ, founded in 1983, and the Christ Apostolic Church, founded in 1985.

1994.⁷ He asked questions about where people currently lived, what ethnic group they belonged to, what their marital status was, what church they were baptized into in Ghana, what church they attend in Philadelphia, whether they were interested in joining a Ghanaian church, and whether they would be interested in joining a subchurch organization such as the choir, women's or men's fellowship, or the BSPG.⁸ His preliminary hunches were substantiated by his survey data. Most Ghanaians did not feel comfortable at American churches and therefore decided not to attend. In fact, 90 percent of Ghanaians surveyed in greater Philadelphia, most of whom were Christian, did not attend any church.⁹ Most respondents said that if a Ghanaian church were established in Philadelphia they would attend.

Apart from the survey, many Ghanaians would attend funeral and memorial services together in Philadelphia. In the summer of 1995, in particular, Ofosu-Donkoh held a memorial service for the late Josephine Mensah at the Elkins Park, Pennsylvania, residence of Louisa Gyandoh, the deceased's sister, and her husband Sam Gyandoh, a Temple University law professor.¹⁰ The large turnout for this funeral by the Ghanaian community in Philadelphia was a significant factor in convincing Ofosu-Donkoh, as well as the Philadelphia Presbytery, that a Ghanaian Presbyterian church was viable in Philadelphia. Ofosu-Donkoh therefore approached the Philadelphia Presbytery with the idea of establishing a Ghanaian Presbyterian Church.

Ofosu-Donkoh presented his survey data to the Presbyterian Church of the USA and suggested that he, an ordained pastor in the Presbyterian Church of Ghana, wanted to establish a Ghanaian Presbyterian Church in Philadelphia. At this time, the Presbyterian Church of the USA had a new church development program that helped to finance emerging churches by putting them on a

---

7. Surveys were distributed at outdoorings, parties, funerals, and African retail stores, as well as mailed to known Ghanaian residents in greater Philadelphia. This information can be found on the UGCC website: http://www.ugccweb.org/pages/about-us/history.htm.

8. Proposed United Ghanaian Community Church, Philadelphia, Preliminary Survey, UGCC folder, Philadelphia Presbytery, Philadelphia. The average age of those surveyed was between twenty-five and thirty-five, and 50 percent were unmarried and without children.

9. Description of New Church Development, the UGCC, UGCC folder, Philadelphia Presbytery, Philadelphia.

10. Ibid.

multiyear budget. Financial assistance decreased over the course of five years as the new church became self-sustaining.[11] The Presbyterian Church of the USA only required that the UGCC follow the Presbyterian Church of the USA organizational structure, which was governed by elders and deacons voted into office by the congregation. All other aspects of church organization and worship were left to the pastor and the congregation.

There were four primary reasons why Ofosu-Donkoh chose to affiliate with the Presbyterian Church of the USA as opposed to the Presbyterian Church of the USA. First, affiliation with the Presbyterian Church of the USA provided financial assistance and access to resources, such as a building for worship and the chance to reserve space at the Stony Point Center in New York. Second, affiliation also meant the founding pastor would not be forced to leave after five years, which is the policy of the Presbyterian Church of Ghana.[12] Presbyterian Church of the USA affiliation meant Ofosu-Donkoh could lead the UGCC indefinitely until he or the congregation decided to break the relationship. Third, the Presbyterian Church of Ghana required larger amounts of money to be donated to the central body than did the Presbyterian Church of the USA. As Ofosu-Donkoh put it, the Presbyterian Church of Ghana wanted money and the Presbyterian Church of the USA provided financial support. Fourth, the Presbyterian Church of the USA had a refugee office near Philadelphia that offered services to UGCC members, such as helping to legalize the status of members, offering legal advice, and lobbying for favorable immigration laws. For these four reasons, Ofosu-Donkoh decided to affiliate his congregation with the Presbyterian Church of the USA as opposed to the Presbyterian Church of Ghana.

Ofosu-Donkoh decided to establish the UGCC as a multidenominational church based on the model of Trinity United Church located on the campus of Trinity Theological Seminary in Legon-Accra, where Ofosu-Donkoh worshipped before enrolling in the seminary. Through his initial survey, Ofosu-Donkoh discovered

---

11. In 1999, the Presbyterian Church of the USA established an organization called the Immigrant Groups Ministries, aimed at providing resources to facilitate the entry of new immigrant groups into the denomination.

12. This issue posed a serious problem for Ofosu-Donkoh. He believed that if he had affiliated with the Presbyterian Church of Ghana, he would have been replaced after five years and forced to take a post in Ghana. This would have impinged upon his goals of accruing wealth and educating his children in the United States.

that the Ghanaian community in Philadelphia represented a variety of denominations other than the Presbyterian Church, such as the Church of Pentecost, the Apostolic Church, the Catholic Church, and the Methodist Church. The UGCC was designed therefore to perform rites in both mainline and Pentecostal fashion: for instance, infant dedication and adult baptism could be performed through immersion for Pentecostals; alternatively, for those from mainline traditions, infant baptism could be performed with water sprinkling. While multidenominational with respect to the performance of church rites, the UGCC was still organized along Presbyterian lines and was meant to be, as Ofosu-Donkoh wrote to the Philadelphia Presbytery, a "local Ghanaian faith community in line with our Presbyterian heritage."[13]

On October 14, 1995, the first abbreviated service was held in Ofosu-Donkoh's apartment, and attendees chose the name United Ghanaian Community Church. While some debate ensued over excluding the word "Presbyterian" from the church's name, the group finally agreed that the exclusion was in line with the church's objective to serve all Christian Ghanaians in greater Philadelphia regardless of denomination. Therefore, the motto of the UGCC became "Serving Christians of All Denominations the Ghanaian Way." This group of ten Ghanaians constituted the founding members and the first interim session or board of elders.

On Christmas Day 1995, the UGCC held its inaugural service at the Holy Trinity-Bethlehem Presbyterian Church, at 1100 West Rockland Street in the Logan section of North Philadelphia, from 2 pm to 5 pm. In attendance that day were approximately 120 worshippers, and the first baptisms were performed. In 1996, the UGCC was admitted into the Presbytery of Philadelphia in the Presbyterian Church of the USA's new church development program. In January of that year, the UGCC claimed 107 total members, which included 43 women, 35 men, and 29 young adults and children.[14] After five years in the new church development program, the UGCC was chartered by the Presbytery of Philadelphia within the Presbyterian Church of the

---

13. Description of New Church Development, the UGCC.

14. Letter, Rev. Ofosu-Donkoh to Rev. Robert Ringstad of the Philadelphia Presbytery, January 15, 1995, UGCC folder, Philadelphia Presbytery, Philadelphia. Rev. Dr. Robert Rigstad, the associate executive presbyter in charge of the Philadelphia Presbytery's new church development program, facilitated the establishment of the UGCC.

USA on January 21, 2001.[15] On that day, Ofosu-Donkoh was officially installed as pastor, and 8 elders and 6 deacons were installed.

Despite UGCC's successful establishment, the church suffered some early attrition of members after the establishment of the Church of Pentecost in Philadelphia later in 1996. After the Church of Pentecost was established, Ofosu-Donkoh claimed that eleven UGCC members left to join it, many of whom had been Church of Pentecost members in Ghana.[16] Of those eleven, five came back to UGCC. Alternatively, the Church of Pentecost pastor in Philadelphia argued that half of UGCC's members switched to the Church of Pentecost after it was established in Philadelphia.[17]

In 2005, the UGCC was the fastest-growing congregation of the 155 congregations that comprise the Philadelphia Presbytery, with a total membership of 198 by December 31, 2005.[18] The UGCC had also become the largest Ghanaian immigrant church in Philadelphia. By June 2006, the UGCC had 222 adult members and 113 children under the age of 18.[19] Nearly 60 percent of all adult members were women and between 30 percent and 50 percent were Presbyterians in Ghana.[20] The great majority of adults (81 percent) were between the ages of 30 and 60; over 35 percent were in their forties.[21] The great majority of adult members were from the Eastern Region or Asante (nearly 85 percent), being ethnically Asante and Akyem primarily,

---

15. The UGCC was the first African immigrant church chartered by the Presbyterian Church of the USA.
16. Author interview with Ofosu-Donkoh, Philadelphia, August 17, 2005.
17. Dittman, "Pentecostal African Church," 23.
18. This information was given during Ofosu-Donkoh's sermon at UGCC on March 12, 2006. This statistic could not be confirmed by the Philadelphia Presbytery.
19. UGCC Membership Roll, June 26, 2006. Alternatively, in the summer of 2005 the Church of Pentecost Philadelphia Assembly had approximately eighty members. Author interview, Church of Pentecost Elder Godfried Asiedu, Philadelphia, August 11, 2005.
20. Of the 220 adult members, 130 were women (59 percent). Only 46 of the 222 adult members had answered the question of previous church affiliation in Ghana. Of the 46, 30 percent had responded "Presbyterian." Just over 20 percent of respondents had previously been Methodist in Ghana. From my own estimates after speaking with members of the UGCC, approximately 50 percent were formally Presbyterian in Ghana.
21. A total of 126 adult members gave their age: 81 percent (102 of 126) were between the ages 30 and 59 in 2006. Children members (under the age of 18) numbered 113 in 2006.

Figure 5.1. United Ghanaian Community Church congregation on January 14, 2007. Photograph by Kwame Ofosu-Donkoh. Reproduced with permission.

and Akwapim and Kwahu secondarily.[22] About one-third of the congregants were middle class by American standards, roughly one half worked low-paying jobs, and another one sixth were students.

The UGCC is currently a member of the Conference of Ghanaian Presbyterian Churches, North America, although Ofosu-Donkoh did not help to establish this organization.[23] It was through this conference's network that Catechist Abboah-Offei came to evangelize at the UGCC, train the UGCC prayer team in Philadelphia, and include

---

22. Comprising the other 15 percent were 15 Fantes, 11 Ewes, 4 Gas, approximately 7 Liberians, an Ivorian who attends infrequently, and a South African.

23. When the UGCC became the first chartered Ghanaian member of the Presbyterian Church of the USA in 2001, the new church development office took the organization off its mailing list. Ofosu-Donkoh was therefore not in contact with the other developing Ghanaian Presbyterian churches that were becoming established within the new church development program. The UGCC did not become a member of the Conference of Ghanaian Presbyterian Churches, North America, or attend the organization's annual meetings until the summer of 2005.

the UGCC prayer team in the New York deliverance workshop. But health and healing was a concern of Ofosu-Donkoh when organizing the UGCC.

Health and Healing in the Early UGCC

If any individual need of the Ghanaian community in Philadelphia was paramount, it was the need for healing: healing of the spirit, the body, and social relationships. Members of the UGCC—like others participating in the New York deliverance workshop (see chapter 4)—continued to be afflicted by malevolent spiritual forces as they had been in Ghana. Ofosu-Donkoh perceived this need, but addressing it took time and finesse, in order avoid alienating older, more conservative church members. Many of these members/leaders were not part of the post-1960s healing movement within the Presbyterian Church of Ghana (see chapter 3).

Ofosu-Donkoh believes that healing—in the broadest sense that includes physical, spiritual, emotional, and social forms—is an essential component of his pastoral duties. In fact, he believes this is mandated by his role as pastor or ɔsɔfo in Twi, the etymology of which points to a healing practitioner. The term ɔsɔfo was originally used to designate a priest who serves local gods for the good of a community, but later came to refer to a Christian pastor or missionary exclusively. In 1881, when the Basel missionary J. G. Christaller first published his Twi dictionary, ɔsɔfo popularly carried both meanings: priest of local deities or Christian pastor.[24]

Christaller traces the etymology of the term ɔsɔfo to the verb *sɔre*, meaning to worship or to perform official religious duties.[25] Ofosu-Donkoh, alternatively, traces the etymology of ɔsɔfo to the verb-phrase sɔ . . . *ano*, literally meaning to seal or stitch the mouth or outlet of something to prevent its contents from leaking out.[26] Sɔ in the context of ɔsɔfo, according to Ofosu-Donkoh, means to stop something misfortunate from continuing or to prevent something bad from happening.[27] Sɔ . . . *ano* is often used in religious contexts, for exam-

---

24. Christaller, *Dictionary*, 467.
25. Ibid., 467, 471.
26. Author interview with Ofosu-Donkoh, Philadelphia, August 9, 2005. Christaller defines this word as to prevent, hinder, or thwart (*Dictionary*, 429).
27. Christaller, *Dictionary*, 429.

ple ɔkomfo Anokye yee anamme de sɔ ɔkom no ano (Prophet Anokye performed customary rites to stop the famine). The primary role of the pastor, argues Ofosu-Donkoh, is to preserve the health and well-being of his or her congregation.[28] Because of this duty, he has always felt the prayer team was an essential component to the life and well-being of his congregation.

In a paper "Mission Design and Goals," given to the Presbyterian Church of the USA when applying to join the organization, Ofosu-Donkoh wrote that "the UGCC seeks to be a congregation that responds through ministry to the spiritual, emotional, and physical needs of people."[29] Those needs were to be met through the establishment of various church groups based on those within the Presbyterian Church of Ghana, particularly the BSPG, or prayer team as it is known among most Ghanaian Presbyterians in North America.

The prayer team was part of UGCC from its founding. It was one of the original organizations listed on the questionnaire that Ofosu-Donkoh used to collect data in the summer of 1994 in order to establish the church. While the UGCC was established in 1996, the prayer team was not instituted until 2000. The founding members of the prayer team were drawn from the worship team, which was formed at the church's inception in 1996. The two teams have different functions and not all prayer-team members are worship-team members. The worship team sings hymns publicly during church services, while the prayer team usually, but not always, functions privately, often in separate rooms throughout the church. The worship team is open to all who are interested in joining, while the prayer team's members are chosen or approved directly by Ofosu-Donkoh.

In 2000, worship-team members Gina Baah, Margaret Smith, and Monica Ofori decided to initiate the prayer team with the approval of Ofosu-Donkoh. The prayer team members met on Wednesday evenings to pray for various concerns, both personal and public, such as health, success, and immigration issues. When Baah graduated from Temple University with a bachelor's degree in business in 2003 and began working, the Wednesday meetings became unsuitable for her, and the prayer team ceased meeting. In early 2005, Ofosu-Donkoh decided to start the prayer team again, with the worship team as its nucleus. Isaac Baah and Rosamund Atta-Fynn were initially added

---

28. Author interview with Ofosu-Donkoh, Philadephia, August 10, 2005.
29. Mission Design and Goal, UGCC File, Philadelphia Presbytery, Philadelphia.

to the worship team of Gina Baah, Margaret Smith, Mary Amoah, Irene Frimpong, and Gloria Baiden. Emmanuel Oppong-Agyare soon joined these seven members. By the summer of 2005, the prayer team was meeting on Friday evenings for three hours of prayer at the home of some of the members in Sharon Hill, Pennsylvania.

Later in 2005, the prayer-team members and some UGCC members—usually between ten and twenty-five people—began gathering in the church sanctuary to pray on Wednesday evenings from 7 to 8 pm. The services began with praying and singing, followed by a short biblical message read aloud by one of the team members. This was followed by the prayer portion of the service during the last twenty minutes, when individual prayer topics were brought forward. The entire group typically would join hands in a circle at the front of the sanctuary, and different people would bring prayer topics forward. Typically these issues revolved around health, marriage, immigration legalities, children's health or success, and pregnancy. Sometimes a person with an especially difficult problem would enter into the middle of the circle, and everyone would pray for that person's particular issue.

Also in 2005, Ofosu Donkoh noticed that during sermons, when he referred to powers of darkness afflicting one's life or God's ability to thwart these powers, people in the congregation responded with various facial gestures that indicated concern, distress, and fear. Ofosu-Donkoh believed that his congregants were certainly thinking of witchcraft or other afflicting agents, even though such thoughts were not voiced publicly in church. These concerns, thought Ofosu-Donkoh, needed to be formally addressed by the church. It was his intention, therefore, to expand the repertoire of the activities performed by the prayer team by developing their role as deliverance practitioners. Catechist Ebenezer Abboah-Offei facilitated this goal.

Ofosu-Donkoh met Abboah-Offei for the first time during the annual meeting of the Conference of Ghanaian Presbyterian Churches, North America, in May 2005, held in Woodbridge, Virginia. Abboah-Offei was present, as was a delegation from the Presbyterian Church of Ghana that included the Presbyterian Church of Ghana moderator, Dr. Rev. Yaw Frimpong-Manso. The Presbyterian Church of Ghana leaders were invited to inaugurate the opening of Ebenezer Presbyterian Church in Woodbridge, Virginia, as part of the National Capital Presbytery. On the final day of the three-day conference, Abboah-Offei approached Ofosu-Donkoh and introduced himself. When Ofosu-Donkoh returned to Philadelphia, he received a phone call from Rev. Samuel Atiemo in Brooklyn—an in-law of Abboah-Offei—who said that Abboah-Offei was coming to the United States on an evangelism trip to raise money for the Presbyterian University in Akropong and wanted to include the UGCC in his itinerary. In late September 2005, Abboah-Offei came to Philadelphia.

Abboah-Offei's First Philadelphia Campaign, September 2005

From September 20 to September 23, 2005, Catechist Ebenezer Abboah-Offei came to the UGCC for a scheduled two-week revival, which was shortened because of a scheduling conflict with another church. Instead, he was in Philadelphia for four days. Flyers distributed by UGCC urged members and nonmembers to come to the services and be "healed, delivered, and anointed" (see fig. 5.2). During this four-day event, Abboah-Offei conducted individual consultations between 12 pm and 6 pm. From 8 pm until 12 am, Abboah-Offei led public deliverance services, where roughly 100 to 120 people were in attendance.

People who came to consult with the catechist arrived in the UGCC sanctuary by 11 am. It was first come, first served. All counselees signed a sheet to indicate the order in which they were to be seen and filled out a deliverance questionnaire. The questionnaire was a modified version of the one used at Grace Presbyterian Church in Akropong. It only asked counselees to write detailed descriptions of their problems. This modified questionnaire was necessary because of the short time Abboah-Offei had in Philadelphia as well as the large number of people wanting to see him. One afternoon I observed fifty to sixty people waiting in the UGCC sanctuary to be seen by the catechist. Loud screams were emanating from one of the rooms where Abboah-Offei was meeting with a counselee. Most of the counselees were not members or regular attendees of the UGCC, and I recognized very few of them. In the evening Abboah-Offei led a public deliverance service.

Standing at the pulpit in the UGCC on the evening of September 20, Abboah-Offei was dressed in a dark suit and had his laptop in front of him with a PowerPoint presentation projected on a large screen behind him. The first scriptural passage that Abboah-Offei referred to was Hebrews 13:8, "Jesus Christ is the same yesterday,

Figure 5.2. Flyer promoting Abboah-Offei's campaign in Philadelphia in September 2005.

today and forever." This biblical quote is meant to demonstrate that the healing activities performed by Jesus in the Bible are possible and even probable for born-again Christians today. Abboah-Offei continued by claiming, "Whatever problem you have, Jesus will solve it for you." This was followed by several testimonies delivered by Abboah-Offei about individuals in Ghana who had been delivered from various ailments and misfortunes after being prayed for by Abboah-Offei or the Grace Team.

The testimonies given by Abboah-Offei demonstrated the range of afflictions that could be healed within this deliverance system. Here, nine exemplars from Abboah-Offei's sermon will be given. One, a man with cancer prayed for his healing, felt gas release from his sides, and afterward physicians could no longer detect the disease. Two, a woman with ovarian cancer was told by her doctor that she could never conceive, but one year after praying with her, Abboah-Offei dedicated her child. Three, a woman came to Grace Presbyterian Church, because her husband wanted a divorce and had stolen her wedding ring. She prayed for her ring to be returned; later, when she went home and cut open a tomato, she found her ring inside of it. Four, the catechist's car was stolen, and prayers were given for its return. The car was recovered the next day, and the thief was apprehended. Five, a blind man was told by doctors that he would never see again, but through prayer his sight was restored. Six, a Pentecostal pastor with a blind daughter was brought to Grace, and after twenty-one days of prayer, her sight was restored. Seven, a man with kidney failure was healed. Eight, a woman came to Grace with her dead baby, and after Abboah-Offei had blessed the baby, it began to shake and was brought back to life. Nine, a former Liberian soldier admitted to killing hundreds of people and, in company with the members of his troop, consuming the organs (particularly the heart) of his enemies.[30] The ex-soldier confessed his sins, became born-again, and was forgiven.

The testimonies of previous cases were repeated for this Philadelphia audience in order to exemplify the broad array of illnesses and misfortunes that can be resolved through Jesus, who continues to have the ability to heal and deliver as in biblical times. After each testimony was given, Abboah-Offei told the audience members that they could attain this same form of divine healing for themselves.

---

30. Stephen Ellis discusses the role of cannibalism, particularly the consumption of an enemy's heart, in the Liberian civil war (*Mask of Anarchy*, 263–65).

Audience members, many of whom had earlier consulted with Abboah-Offei and were healed, gave several healing testimonies during these services. Others gave healing testimonies that demonstrated the power of prayer generally. During the evening service on September 21, Pastor Ofosu-Donkoh gave his own healing testimony. He said he had had high blood pressure for four years, and his physician tried to control the disorder by restricting his diet. His blood pressure was typically around 140/100, whereas the norm is 120/80. His doctor had recently suggested that he take blood pressure medication at a cost of sixty dollars per bottle, which he would have to take three times a day. He prayed about this matter and went for his final blood pressure checkup before the medication was prescribed. The doctor checked his pressure, and miraculously it was 120/80 for the first time in four years.

Besides the plethora of healing testimonies given, Abboah-Offei also discerned audience member's afflictions by divine revelation. During his services, and often in the time between giving testimonies, Abboah-Offei would ask an audience member to come forward, and would diagnose a disorder and pray for them. Other times the catechist would say that someone in the audience had some specific disorder, look around the room, and then call someone out of the audience to be prayed for. The catechist would touch these people brought forward during public service on the head, wrists, or shoulder, or else clasp their hands. Many of those called forward became possessed. It was this ability to diagnose afflictions revealed by God that most informants agreed was the most important and powerful aspect of Abboah-Offei's healing ministry.

Another memorable case from Abboah-Offei's 2005 evangelism trip to Philadelphia, recalled by Ofosu-Donkoh, was a man delivered from evil spirits that inhibited his ability to learn.[31] The man went through ten years of formal education but was not able to write anything, not even his own name. The catechist deduced that this demonic oppression was caused by a family member in Ghana cursing or bewitching him. They prayed against those demonic forces in his family, and he was delivered. Afterward, he became successful in school.

This evangelism trip by Abboah-Offei caused two significant changes in the prayer team in Philadelphia. First, the prayer team

---

31. Author interview with Ofosu-Donkoh, Philadelphia, October 10, 2005.

started learning and preparing to counsel the afflicted using Abboah-Offei's deliverance materials. Second, this trip also prompted the team's growth and strengthened the commitment of its members. In November 2005, five members of the Philadelphia prayer team went to the Stony Point Retreat Center in New York for the week-long deliverance workshop taught by Abboah-Offei. In early 2006, there were eight members of the UGCC prayer team, which included Isaac and Gina Baah (a married couple), Mary Amoah and Irene Frimpong (sisters), Rosamond Atta-Fynn, Gloria Baden, Margaret Smith, and Emmanuel Oppong-Agyare. This prayer team was growing in strength and responsibility, and its legitimacy increased further in the course of and following Abboah-Offei's second Philadelphia evangelistic campaign in April 2006.

Abboah-Offei's Second Philadelphia Campaign, April 2006

Abboah-Offei visited the UGCC from April 24 through April 30, 2006. His schedule was fairly similar to the previous visit. Abboah-Offei held public deliverance services on Monday through Saturday evenings from 9 pm to 12 am and Sunday from 11 am to 2 pm, which averaged between 100 and 120 attendees per evening. On Monday, Tuesday, Wednesday, and Friday, Abboah-Offei met with afflicted individuals for consultations between 12 pm and 8 pm, using the modified deliverance questionnaire.

The major difference between this campaign and his previous campaign was the formal teaching of deliverance practices to the Philadelphia prayer team, which occurred between 11 am and 6 pm on Thursday and Saturday during that week. During these teaching sessions, Abboah-Offei lectured for two three-hour stretches (with a one-hour break) about the theory of deliverance, and various cases were presented to demonstrate proper counseling of afflicted individuals. No one else gave lectures, and there was no music or dancing. All participants met in a room on the second floor of the church.

Abboah-Offei began the two-day seminar by discussing the difficulties of the deliverance ministry. The ministry is confrontational, argued Abboah-Offei, and counselors fight evil spiritual forces. Deliverance counselors must have a passion for the ministry and must abstain from sinful practices such as alcohol consumption, infidelity, and idolatry.

After discussing the qualifications of a deliverance practitioner, he reviewed cases from Grace Presbyterian Church. He emphasized that spiritually afflicting agents existed not only in Africa, but were also prevalent in the United States. Abboah-Offei discussed followers of American witchcraft (Wiccan) as well as West Indian followers of African-derived religions such as Vodou and Santeria, as examples. These people, and the spirits whom they serve, were all potential sources of spiritual harm.

In particular, Abboah-Offei focused on two types of agents that afflicted the Ghanaian diaspora community: witches and spirits invoked in Akan therapy. For Abboah-Offei, witchcraft was a universal cross-cultural phenomenon, as well as the most common type of afflicting agent in Ghana and the diaspora. Another common cause of affliction that emerged in adults could be traced to their receiving protection from deities as children through a process of facial incisions (ɔkam) with inserted pharmacopoeia (mmoto), performed by prophets (ɔkomfo).[32] Survey data collected by Abboah-Offei among counselees over forty-five years of age at Grace Presbyterian Church showed that 82 percent had gotten as a child some form of incision on their bodies by an Akan healer for the purpose of protection against malevolent spiritual forces.[33] The result of the spiritual bond formed between those children and the various spiritual entities caused Ghanaians at home and in the diaspora illness and misfortune. The scars that resulted from these facial incisions were evident on the cheeks of many members of the UGCC.

Besides forms of spiritual affliction and Akan therapeutics, biomedicine was discussed. Abboah-Offei described the integration of biomedicine in his ministry at Grace, particularly the contributions of doctors and psychiatrists. He stated during an interview that he wanted to establish a healing center with intercessors and deliverance counselors working cooperatively with doctors, nurses, and psychiatrists equipped with modern hospital technology.[34]

---

32. Abboah-Offei's deliverance questionnaire asks counselees whether they have visited a native doctor, whether they have had incisions made on any part of their body, and whether they have ever taken any black powder (mmoto) (Abboah-Offei, "Church Leaders," 20).

33. Lecture given by Abboah-Offei at the New York deliverance workshop, November 26, 2007.

34. Author interview with Abboah-Offei, Akropong, February 15, 2007.

There were moments during which areas of conflict between the deliverance ministry and biomedicine were evident, however. Abboah-Offei shared stories about physicians who were skeptical of the results at Grace or challenged the legitimacy of Christian healing. He also shared instances in which people solely committed to biomedicine threatened to sue him. He continued by speaking about the necessity of biomedicine, although he claimed that spiritual disorders managed through prayer were resolved much faster than disorders cured through biomedicine.

Psychiatric disorders were also discussed. Their relation to demonic possession needed clarification for the Philadelphia prayer-team members. Abboah-Offei differentiated between what he called "demonic schizophrenia" and "psychological schizophrenia," which present slightly different symptoms. Demonic cases hear voices within their heads, while psychological cases hear voices behind them. Furthermore, the catechist suggested that multiple personality disorders and demonic possession were nearly the same thing. In some cases multiple personalities could be distinguished from demonic possession by observing eye movements. In cases of multiple personalities the eyes of the afflicted tend to blink rapidly, while in cases of demonic possession the sufferer's eyes tend not to blink at all. A common method for diagnosing demonic possession cases is through prayer, that is, diagnosis via therapy.[35] If a counselor prays against malevolent spiritual forces with the afflicted person and he or she begins to show physical and verbal signs of possession (i.e., manifestation), then the cause is demonic.

As in his deliverance workshop in New York, there was also a practical component in which the afflicted were brought in and prayed over in order to give the Philadelphia prayer-team members hands-on experience managing cases of illness and misfortune. One man brought his brother from Florida to be prayed for by the catechist. The man and his sick brother, who had an advanced form of cancer, entered the room. The catechist spoke to the sick brother, prayed with him, and confirmed that he was born-again. Abboah-Offei laid hands on the man's face, which had visible cancerous tumors on it, and vigorously prayed for "God to smite the dirty cells," which originated with the devil. Spirit possession did not ensue and there was no diagnosis of a malevolent spiritual force behind his disease. The

---

35. Feierman, "Therapy as a System-in-Action."

catechist simply prayed for divine intervention for healing. Other cases, however, brought before the prayer team by Abboah-Offei did result in spirit possession. With all these cases, the Philadelphia prayer team got to observe, diagnosis, and pray with those suffering from various afflictions.

In Abboah-Offei's April 2006 trip to Philadelphia, he contributed to the formation of the prayer team through the training of deliverance counselors. He also served the needs of the afflicted individuals by performing individual consultations. The sermons given by Abboah-Offei during the evening also helped give a public voice to the struggles and afflictions of Ghanaians in the diaspora. These deliverance sermons or lectures, performed publically (as opposed to individual consultations), incorporated the Philadelphia prayer team, demonstrating their abilities to the entire congregation.

Abboah-Offei gave an integrated set of sermons during his April 2006 visit to Philadelphia, addressing migration, afflictions, and healing. He did not simply preach about the healing power of Jesus and the range of disorders treatable through faith in God, as he had during his 2005 visit. Abboah-Offei used the biblical story of the Israelite migration out of Egypt to the Promised Land to exemplify how those migrants, through faith, prayer, and the observance of God's laws, overcame hardships in order to find prosperity in a new land. The theme of this presentation was "crossing over." As the Israelites had crossed over from slavery to prosperity through their migration, the catechist urged Ghanaian Presbyterian immigrants in Philadelphia to cross over from demonic affliction to health and prosperity in their new land. Abboah-Offei asked, "Is your marriage or move to the United States not what you expected? The people thought they left Ramses for the Promised Land . . . but went through to Sukkoth and they starved! The faith in God of the Israelites is what brought them to the Promised Land." Abboah-Offei exhorted the Ghanaian community to stay faithful to God, so that they might receive their rewards in Philadelphia as the Israelites had in the land of Canaan.

At times, the analogy between the suffering of the Israelites and the Ghanaians in their migrations would fuse within Abboah-Offei's sermons, collapsing the distinction between biblical text and ritual context.[36] For example, Abboah-Offei stated, "Because you are in the

---

36. See Keane, "Religious Language," 60.

desert, the government is robbing you because you don't have the paper!"[37] For him, Israelites suffering in the dessert and Ghanaians suffering from their illegal status (referred to as "paper," i.e., citizenship papers) was conjoined in the pronoun "you," the audience. By the discursive collapsing of space and time within this text segment, the Israelite migrants and Ghanaian migrants became rhetorically equivalent. Therefore, the structure of this utterance is related to its ritual action;[38] it defines the struggles of this migrant community as one of God's chosen people. The Ghanaian struggle was not just material, but also spiritual and divine, as they had become God's chosen people through Abboah-Offei's sermon.

Abboah-Offei continued the analogy through the course of his week-long presentation, particularly focusing his attention on the topic of faith in God, which for Abboah-Offei largely meant abstaining from sinful behavior. The Israelites spent forty years in the wilderness before reaching the Promised Land because of their sinful behavior, and many Ghanaians in the United States were slow to become successful because of sin. In terms of sinful behavior, the catechist leveled considerable blame on those who consulted alternative religious or therapeutic practitioners for protection or healing. The Israelites at one point had lost faith in God and began worshipping idols (the golden calf), just as the Ghanaian community relies on shrines or charms for protection and success in their new environment. Abboah-Offei ultimately argued that the Ghanaian community should have faith in God, live free of sin, and rely on the prayer team to manage their health and welfare.

During these public services in 2006, members of the Philadelphia prayer team participated in handling the deliverance cases that emerged during each service. Some people voluntarily came forward; others were called forward by Abboah-Offei (via discernment by divine revelation) and were prayed over. Many became possessed and were assisted by the prayer team, who restrained, comforted, and covered them as well as cleaned up their vomit. This process demonstrated the prayer team's ability to deliver afflicted persons, as well as its legitimate right to do so as protégés of Catechist Abboah-Offei.

---

37. Abboah-Offei Sermon at the UGCC in Philadelphia, April 25, 2006.
38. See Silverstein, "Metaforces of Power."

The UGCC Prayer Team from Prayer Group to Deliverance Counselors

The two evangelism trips by Abboah-Offei to Philadelphia in 2005 and 2006, as well as the prayer team's participation in the 2005 deliverance workshop in New York, shifted the focus of the prayer team. Instead of solely praying about issues in the church, the team began to treat afflicted individuals with various problems by applying diagnostic techniques established at Grace Presbyterian Church. After Abboah-Offei's visit to the UGCC in 2006, Isaac Baah, Rosamund Atta-Fynn, and Emmanuel Oppong-Agyare began meeting with people in private rooms in the UGCC during either Sunday services or Wednesday prayer services for deliverance. This happened initially with afflicted individuals diagnosed by Abboah-Offei in Philadelphia during his 2006 campaign.

In the six-month period after Abboah-Offei's 2006 campaign, the prayer team handled six to seven cases within the church. Some of these cases were seen by individual prayer-team members through individual consultations, while others were seen by part of or the entire prayer team. Nearly half of these cases were people who had come to see Abboah-Offei in April 2006 and whose cases were handed over to the Philadelphia prayer team. The prayer team was in the process of systematizing their meetings, practices, and diagnostic techniques, for example having counselees fill out deliverance questionnaires. After the catechist's two visits, the prayer team continued to hold Wednesday hour-long services from 7 to 8 pm.

By the end of 2006, the prayer team began to play a more public role in the church. On October 29, 2006, Ofosu-Donkoh called the prayer team forward to publicly pray for the church and various afflicted individuals during a Sunday service. This was the first time that the group was formally introduced to the church during a Sunday service. During this service, Ofosu-Donkoh declared, "There is power in the hands of Jesus. Open up and you will discover that he is a healer. You will go to the doctor. You will go to the dentist. And the doctor only treats the sicknesses. But if you let the cloak off and you open up to Jesus, you will find not only a doctor but you will find a healer." If you have any problem, urged Ofosu-Donkoh, bring it before the church and be prayed for. The prayer team came forward and prayed for people in the church, laying hands on some. There were no cases of possession, but it was the first time that the prayer team was incorporated into a public Sunday service.

A month later, during the last week of November 2006, the Philadelphia prayer team participated in their second deliverance workshop in New York. Participants from the UGCC prayer team included Mary Amoah, Irene Frimpong, Gina Baah, Isaac Baah, Rosamund Atta-Fynn, Emmanuel Oppong-Agyare, and David Frimpong (pseudonym), who was previously delivered from a spiritual affliction (see case study in chapter 6). The Philadelphia prayer team played an especially noticeable role compared with other prayer teams. Mary Amoah, Irene Frimpong, and Gina Baah, worship-team leaders in the UGCC, led the singing during the workshop. Rosamund Atta-Fynn gave a powerful sermon on the morning of November 28, 2006. Isaac Baah and Emmanuel Oppong-Agyare worked in conjunction with Samuel Asare (Abboah-Offei's assistant) as deliverance counselors during public services, tending to those who became possessed. Also, Isaac Baah did some video recording for Abboah-Offei, since he was not present and wanted to observe what was happening at the workshop.

In early 2007, the Philadelphia prayer team began meeting once a month to study Abboah-Offei's deliverance manual and improve their skills as deliverance counselors. By the summer of 2007, the prayer team began meeting every Monday evening to study Abboah-Offei's deliverance manual. Part of what they discussed was their own spiritual exposures and ways to overcome them, so as to free themselves from any demonic power and thus become more productive deliverance counselors. Also in 2007, the prayer team began to lead one Sunday service every three months that resembled their Wednesday prayer services, but included the entire congregation. Beginning in 2008, Pastor Ofosu-Donkoh planned to transform these Sunday prayer services into deliverance services, similar to the services performed by Abboah-Offei on his previous visits to Philadelphia.

The Wednesday prayer service is still the primary occasion when the Philadelphia prayer team acts publicly. During a Wednesday service in early 2007, one of the UGCC members acknowledged her suffering to the group, was prayed for, and became possessed. Her name was Ama Asare (pseudonym), and she was the first counselee to go through the deliverance process exclusively under the direction of the UGCC prayer team.

The UGCC Prayer Team's First Case, Ama Asare

After Abboah-Offei's 2006 campaign, the Philadelphia prayer team followed up several cases that were originally diagnosed by Abboah-Offei and began counseling newly afflicted members. The first case seen solely by the UGCC was a woman named Ama Asare, who had struggled with an alcoholic husband in Ghana, suffered from a variety of bodily pains, consulted several types of healers in Ghana, and had an affair with a man diagnosed with HIV in the United States.

Let me recount Asare's case—a narrative of illness that spans several years and two continents—as it was told to me as well as to the UGCC prayer team. Asare's mother is Ga and her father Akwapim (from Aburi). She grew up in Tema, a coastal town in greater Accra, and was active in the Methodist Church as a child but later joined the Anglican Church when she enrolled in an Anglican secondary school. In 1982, Asare was married and moved to her husband's village, where she began working as a petty trader. Soon afterward, she began to experience a sharp pain in her chest and side that she described as feeling like a bullet was passing through her. Asare believed that someone, possibly an envious person or competitor in the marketplace, had cast a spell or bewitched her.

Asare's physical problems were compounded by marital difficulties. Soon after Asare was married, her husband started smoking, drinking, and acting immorally. This behavior was troubling to Asare, who felt she must act on behalf of her marriage. Different friends and family suggested that she take her husband to various therapeutic practitioners in order to treat his addiction and antisocial behavior. Initially, she decided to take her husband to a prophetess of an Aladura church.[39] Both Asare and her husband attended the Aladura church for a while, but soon afterward her husband stopped going, while she continued. Without her husband's participation in the therapy, there was no change in his behavior, although some of her bodily pain did dissipate.

After six months, she took her husband to a prophet (ɔkɔmfo) for help with his condition. The prophet told them that if they followed

---

39. This church was probably a branch of the Church of the Lord (Aladura), an independent church established in Nigeria in the 1930s that later spread into Ghana. Of note, many of the Nigerian founders of this church were members of Faith Tabernacle, reading their literature, or in communication with the Philadelphia leadership in the 1920s (Turner, *History of an African Independent Church*).

the guidelines set forth by the deity, her husband's problems would be resolved. Various rites were performed in order to create a covenant with the deity. In particular, Asare and her husband had their heads shaved and were bathed in goat's blood over the course of three days. After these events, they were told to return weekly for successive rites. After two or three weeks, Asare's husband stopped going. She tried forcing him to go, but he refused. Asare continued visiting the shrine for a while, but eventually decided to stop because her husband was not following the prescribed set of duties required to serve that particular deity. Eventually Asare broke ties with the prophet by presenting her with some cola nuts and a rooster.

After this failed attempt, Asare decided to consult her husband's family about his problem. Asare's eldest brother-in-law suggested another prophet. She went with her brother-in-law to consult this prophet, but her husband refused to come. She presented the prophet with some gifts initially, but only consulted with this practitioner once. Asare's husband's problems were never resolved. Eventually Asare immigrated to Philadelphia in 2004, without her husband, to stay with her sister.

Soon after her immigration, she and her sister joined the UGCC. Asare also began dating a local man, but soon thereafter he was diagnosed with HIV. Asare was scared that she would become infected and die. So she approached Gina Baah after a service in 2004, told her the problem, and began to cry. Gina called Isaac Baah, her husband, and Emmanuel Oppong-Agyare, and they all prayed together. Furthermore, they fasted with her for one week. The prayer team at that time prayed for her during a Wednesday prayer service and advised her to get tested, which she did. The test came back negative.

After Asare got the test results, she discontinued coming to prayer services; in 2004 the prayer team was not trained in deliverance and Asare felt they could do little else for her. She did not come for consultation during either of Abboah-Offei's visits to Philadelphia in September 2005 or April 2006. While she did attend one of the catechist's public deliverance services at UGCC in April 2006, she did not come for consultation until her health started deteriorating in 2007.

In early 2007 Asare had not been diagnosed with HIV, but she experienced unexplained joint pain and a frequent burning sensation in her back and legs. She also began losing weight, which caused her much distress. She went to the hospital twice to inquire about the source of her bodily pains, but no diagnosis could be rendered. Because Asare was in the United States illegally, her sister felt visiting the hospital would jeopardize her ability to stay and advised her not to go again.

In March 2007 Asare went to a Wednesday Prayer Meeting. She presented her problem before the prayer team, who prayed for her. Asare became possessed, coughed violently, and vomited. They encouraged her to confess her sins, which included consulting an Aladura priestess and two prophets, as well as committing adultery. After this event, the prayer team gave her a deliverance questionnaire to fill out. She did not complete the form entirely because she did not understand all the questions.[40] The prayer team continued praying and fasting for her.

Isaac Baah described Asare's visit to the two prophets as the most potentially dangerous thing she had done. It was the covenant made between her and the deity that was probably the primary afflicting agent in her case. The Aladura Church is potentially dangerous, but in this case not as harmful. Finally, committing adultery is a common way to become demonically possessed or oppressed, and Baah believed that this was also a problem for her. All three activities—visiting the Aladura priestess, consulting the two prophets, and committing adultery—were believed to be causing her demonic possession and afflicting her with bodily pains. There was no single cause.

By late 2008 Asare was still suffering. In November of that year, Asare returned to UGCC for a public deliverance service led by Abboah-Offei. After the sermon, Abboah-Offei and other prayer-team members invited congregants to come to the front of the church and be prayed for. Among the many people who came forward was Ama Asare.

When Asare joined hands with Catechist Abboah-Offei, she was visibly crying. Soon after Abboah-Offei began praying with her, she collapsed and began shouting; Asare was spirit-possessed. Abboah-Offei continued to grasp her hands and pray with her, until several prayer-team members, led by Emmanuel Oppong-Agyare, led her to the back of the church for further therapy. Oppong-Agyare did not lead her into one of the back rooms, where most spirit-possessed congregants are taken when possession occurs, but continued to pray for Asare in the hallway that connects the sanctuary to the back rooms. Oppong-Agyare, assisted by two other prayer-team members,

---

40. This is a common problem at Grace Presbyterian Church; therefore, a group of women on the Grace Team assist the afflicted in filling out the questionnaire, explaining each question to them.

furiously commanded the demonic spirits to part from Asare, who at this time was sitting in a chair. Asare continued crying and screaming, and vomited once. Periodically, her whole body would convulse, while Oppong-Agyare stood in front of her forcefully praying. She seemed to be in a semiconscious state.

Ama Asare's case was significantly complex. It involved several types of afflicting agents (witches or jealous coworkers, traditional healers, alternative Christian healers deemed illegitimate), both material and spiritual afflictions (pains in chest, sides, and joints, burning sensations in the back and legs, as well as spirit possession and curses), immoral behavior (an affair), as well as fear and insecurity (potentially HIV positive, illegal immigration status). As of this writing, Asare has not experienced a significant transformation, as have some other cases in the UGCC (for instance, see the case of David Frimpong in the following chapter). This first case did, however, give the prayer team the experience and confidence to minister to the afflicted without the assistance of the catechist Abboah-Offei. To date, however, some Ghanaians in Philadelphia still initially contact Abboah-Offei in Akropong, but are redirected to the UGCC prayer team by the catechist for healing and deliverance.

Healing in Communion Prayers

While the prayer team is the primary group responsible for managing the spiritual afflictions of church members, it does not have a monopoly over discourses of healing within UGCC. Healing and deliverance are frequent topics of exhortation during Sunday sermons presented by Ofosu-Donkoh. But in particular, healing powers are addressed during the invitation prayers for communion, in which the congregation is invited to take part in the ritual of consuming bread and wine consecrated as the body and the blood of Jesus Christ. Public healing within communion prayers demonstrates the importance and necessity of healing for all believers, which in turn can be met more directly through deliverance practices offered by the prayer team.

Communion, along with baptism in the Presbyterian tradition, is one of two sacraments: rites believed to be instituted by God and commanded by Jesus for his followers to perform. Communion in the Presbyterian tradition is much less formal than the communion rite in the Lutheran or Anglican churches. In preparation for communion with Jesus, believers must confess their sins, reconcile with God and all people, and trust in Jesus Christ for cleansing and renewal.[41]

The Presbyterian communion is a predetermined set of utterances, referred to as the liturgy, shared by all Presbyterian churches, although differing slightly between nationally based Presbyterian churches, such as the Presbyterian Church of the USA and the Presbyterian Church of Ghana. Often the liturgies have several prayers or utterances that the pastor can choose from. In the UGCC, however, Rev. Ofosu-Donkoh has taken the liberty of amending various standardized prayers read during communion. By and large these amended prayers have included references to healing, which is common among Pentecostals[42] and many Eastern Orthodox,[43] for whom communion is often depicted as a form of medicine.

Communion in the UGCC, through the prayer of invitation primarily, has become a rite of cleansing, renewal, and healing. For example, on September 7, 2005, Ofosu-Donkoh's invitational prayer stressed healing: "Transform yourself through communion. Transform the communion into medicine. Drink and be healed. Don't need to sacrifice a lamb or a cow or a chicken. The blood of Jesus is the only and ultimate sacrifice. If you are sick, let the body and blood be your medication." Ofosu-Donkoh's reference to sacrificing animals is intentionally inserted to juxtapose the healing powers of Christianity to the healing powers of Akan therapeutic practitioners, who often require the sacrifice of animals in order to receive healing. Ofosu-Donkoh emphatically stated that it is Jesus' blood, not the blood of sacrificed animals, that ultimately heals. This points to an important component of deliverance theology, particularly espoused by Abboah-Offei, that consulting Akan healers is a spiritually dangerous and ultimately harmful endeavor.

From July 2005 through July 2007, in communion prayers and particularly during the invitation prayer, an invocation of God's healing powers was a common occurrence. The healing power did not solely refer to physical ailments, but also to spiritual problems, demonic possession, antisocial disorders, fighting between family and friends, joblessness, barrenness, protection and success for children,

---

41. Presbyterian Church of the USA, "Presbyterian 101," 3.
42. Karkkainen, "Pentecostal View," 126–28.
43. Harakas, *Health and Medicine*, 91.

immigration problems, and illnesses incurable by biomedicine.[44] Discourses of healing found in these communion prayers addressed misfortune as well as threats to survival and reproduction.

The importance of this invocation, during the one rite believed to bring all baptized members into closest communion with God, demonstrates the importance of healing in the life of the UGCC and its members. Publicly, healing can occur through communion. Privately, the afflicted at the UGCC have the opportunity to consult the prayer team for healing and deliverance, although increasingly the prayer team has also taken a more public role. Their more recent public presence in the UGCC is due to all prayer-team members being voted into leadership positions as elders and deacons in December 2006.

The UGCC Prayer Team Members Become Church Leaders

Many of the oldest and earliest members of the UGCC, who initially held positions as elders and deacons from 1996 through 2006, were opposed to or uncomfortable with healing and deliverance practices. These members grew up within more conservative mainline churches, such as the Presbyterian Church of Ghana, in which healing and other charismatic practices were absent. For many of these individuals, the absence of these practices is what separated the Presbyterian faith from Pentecostalism. Few of them had experience with Scripture Union, the BSPG, or other Pentecostal churches in Ghana. Most were chosen as leaders because they were the oldest members, following a Ghanaian pattern of choosing leadership based on age. Some members of the Board of Session particularly disagreed and opposed Ofosu-Donkoh's invitation to Abboah-Offei for his two campaigns in Philadelphia in September 2005 and April 2006.[45] They believed that the UGCC was already becoming too Pentecostal. But Ofosu-Donkoh made the final

---

44. See communion prayers by Ofosu-Donkoh on August 21, 2005; October 2, 2005; January 1, 2006; February 5, 2006; April 2, 2006; August 6, 2006; October 1, 2006; December 3, 2006; June 3, 2007.

45. Members of the Board of Session complained that many nonmembers participated during Abboah-Offei's September visit without contributing to the fees paid for his services. Alternatively, Ofosu-Donkoh believed participation was good exposure for UGCC. He had not charged nonmembers at the door, as the Board suggested, because Ofosu-Donkoh believed that doing so would not have been good church politics.

decision, overruling the Board of Session in bringing Abboah-Offei to the UGCC.

After Abboah-Offei's initial campaign, the Board of Session would not fully sanction the prayer team to perform individual consultations with church members.[46] They were concerned about possible litigation in case someone was injured. To appease them, Pastor Ofosu-Donkoh argued that he would always be present during these counseling sessions, and since he, as pastor of the church, had a legal right to counsel members, in the broadest sense, it would prevent possible litigation. As a group, they were opposed to the prayer team, an organization that Ofosu-Donkoh believed to be essential for the success of the UGCC.

Ofosu-Donkoh and some members of the church leadership were at odds on the issue of spirituality and spiritual healing in the church. Ofosu-Donkoh wanted leaders with high spiritual qualities to lead the UGCC, not just the oldest members.[47] In mid-December 2006, the UGCC held elections for elders and deacons, and before the nominations took place, Ofosu-Donkoh spent a significant amount of time teaching and encouraging the UGCC members to choose leaders with qualities similar to the prayer-team members and biblically mandated in Titus 1:6–9 and 1 Timothy 3:1–23.

As early as November 12, Ofosu-Donkoh announced the coming election of elders and deacons, and every week thereafter discussed their necessary qualities: abstaining from alcohol, conducting themselves properly in public, being monogamously married or single, and being on good terms with all members in the church. Ofosu-Donkoh even handed out literature describing how elders and deacons are responsible for the spiritual development and pastoral care of the church and its members. The literature again listed qualifications of both offices, which matched those of the prayer-team members.

The link between the upcoming vote for elders and deacons and the prayer team was particularly evident during the Sunday service on December 3, 2006, when a detailed description of the qualities of deacons and elders was again stated by Ofosu-Donkoh just after a report by Atta-Fynn about the prayer team's participation in the deliverance workshop in New York. The pairing of the two speeches by Ofosu-Donkoh and Rosamund

---

46. Author interview with Isaac Baah, Philadelphia, September 10, 2006. Author interview with Ofosu-Donkoh, Philadelphia, December 16, 2007.

47. Also, none of the members of the Board of Session were elders or deacons in their previous churches in Ghana and therefore had no experience governing a church.

Atta-Fynn drove home the point that particularly the prayer-team members were those with the qualities befitting deacons and elders.

On December 17, 2006, the UGCC held formal elections in which nine elders and nine deacons were elected in the official Presbyterian format. None of the previous elders or deacons were voted into office, as they had already served their terms and were not eligible for reelection. All nine prayer-team members were voted into the eighteen leadership positions. As a voting bloc supported by Pastor Ofosu-Donkoh, the prayer team secured a majority. Ofosu-Donkoh was hopeful that new leadership would result in more spiritual development, healing, and deliverance within UGCC. On January 14, 2007, the new UGCC elders and deacons were installed through a ritual of laying on of hands by Rev. Ofosu-Donkoh and other leaders from the Philadelphia Presbytery, which was followed by a celebration (see fig. 5.3). As the primary voting bloc within the UGCC polity, the prayer team had the capacity to further integrate itself and offer deliverance services to the church's members.[48]

Conclusion: Philadelphia's Enchanted Calvinism

The UGCC prayer team was initially little more than an idea of Dr. Rev. Kobina Ofosu-Donkoh, the founding pastor of the UGCC, who thought of creating a Ghanaian Presbyterian Church in Philadelphia and replicating various church groups, such as the BSPG, within his congregation. A few years after the church's establishment in 1996, the prayer team became a small group that gathered together to pray, but as members' schedules became too hectic, it ceased to function before reemerging in early 2005.

From the time that Catechist Ebenezer Abboah-Offei led his first campaign at the UGCC in September 2005 through the Philadelphia prayer team's participation in its fourth deliverance workshop in November 2008, the UGCC prayer team changed significantly. Through the interaction and guidance of Catechist Abboah-Offei, the premier deliverance practitioner of the Presbyterian Church of Ghana, the UGCC prayer team was trained systematically to replicate the healing practices implemented at Abboah-Offei's church in

---

48. Among other things, the prayer team is now funded to go to the annual deliverance workshop in New York (which in 2008 cost $600 per person for the week), as well as perform deliverance services more frequently during Sunday services.

Figure 5.3. Celebration at the United Ghanaian Community Church after the installation of elders and deacons on January 14, 2007. Photograph by Kwame Ofosu-Donkoh. Reproduced with permission.

Akropong, Ghana. The prayer team was transformed into a group of deliverance counselors who diagnosed and managed the welfare of afflicted individuals in the UGCC congregation. Finally, the prayer team was voted into the church's two political offices, which cemented the centrality of healing and deliverance within the congregation. Through the process of institutionalizing healing and deliverance practices within the UGCC, this Philadelphia church became enchanted. Spiritual explanations of illness and misfortune became publically recognized, malevolent spiritual forces such as witches were determined to be the culprits, and the prayer team managed the healing process to combat these malevolent forces.

Chapters 4 and 5 demonstrated how deliverance practices were replicated in North America among the network of Ghanaian Presbyterian churches as well as within one particular congregation, the UGCC in Philadelphia. In the following chapter, I will examine how and why deliverance disjunctures have occurred in the Ghanaian Presbyterian diaspora, focusing specifically on the transformation of gendered deliverance practices.

6

Gendered Transformations of Enchanted Calvinism in the Ghanaian Presbyterian Diaspora

This chapter is different in its scope than the previous five. While the prior five chapters are presented in chronological sequence, from the early nineteenth century to the (ethnographic) present, chapter 6 encompasses this entire time frame. While chapters 1–5 focus on *how* the Presbyterian Church of Ghana became enchanted in both Ghana and North America, chapter 6 takes enchanted Calvinism as a premise. Chapter 6 answers the questions *what* and *why*, explaining what form enchanted Calvinism takes in the United States and why it has taken this form.

More specifically, chapter 6 focuses on gendered transformations of enchanted Calvinism in the Ghanaian Presbyterian diaspora. At Grace Presbyterian Church, the deliverance practitioners are primarily men, and the spirit-possessed are exclusively women. In the Ghanaian Presbyterian diaspora, however, women have also become deliverance practitioners and some men have become spirit-possessed.[1] Spirit possession is a phenomenon that entails spiritual agent(s) taking over a host's executive control or replacing a host's mind, therefore assuming control of a host's bodily behaviors and utterances.[2] Spirit possession is one of the two primary ways that

---

1. Data presented in this chapter were derived from ethnographic research conducted with the UGCC in Philadelphia from 2005 through 2009, at the New York deliverance workshop from 2006 through 2008, among various other participants from Ghanaian Presbyterian churches in North America, and finally, at Grace Presbyterian Church in Akropong in February 2007.

2. See Cohen, "What Is Spirit Possession."

Satan can affect people, which in turn requires deliverance in Ghanaian Christianity.[3] The other is spiritual affliction, where illness or misfortune is attributed to Satan. Within the Ghanaian Presbyterian community, spirit possession is also pathogenic or harmful and thereby integrally related to spiritual affliction. Because of this close relationship between possession and affliction, spiritual possession in the context of this community refers exclusively to executive possession by malevolent forces and *not* the Holy Spirit, which also has the capacity to take executive control of a host. Possession by the Holy Spirit is never described as "spiritual possession," but rather as "baptism with/in the Holy Spirit" or as "slain by the Spirit." The goal of deliverance is for the malevolent spiritual agents to be driven from their host and replaced with the Holy Spirit, which is benevolent and beneficial.

Among Ghanaian Presbyterians in the United States, this recent transition toward gender egalitarianism in Christian healing rituals involving spirit possession is not simply another chapter in the "sex wars" described by I. M. Lewis.[4] Lewis explained possession as a thinly disguised protest movement directed against spouses; in most cases wives became ill and were not cured until their possessing spirits' requests (which mask wives' requests) for material gains were met by the husbands.

There are two problems with Lewis's instrumentalist understanding of spirit possession, one specific to Christian therapy and the other more theoretically broad. First, possessing spirits are never to be appeased in Christian spirit possession generally, and specifically among Ghanaian Presbyterians, unlike in *Zar* cults in Somalia and Sudan. Christian sufferers never willingly engage in a productive relationship with these spirits, and subsequently, requests made by spirits during the course of spirit possession are never placated. The goal of Christian deliverance is not to pacify possessing spirits but rather to exorcise and replace them with the Holy Spirit. Possession by malevolent spirits in this context cannot provide a direct means for the gratification of wishes ordinarily denied to the possessed.[5]

Second and more broadly theoretical, spirit possession has much greater articulatory potential than Lewis's instrumentalist

---

3. Asamoah-Gyadu, *African Charismatics*, 167.
4. Lewis, *Ecstatic Religion*.
5. See Bourguignon, "Suffering and Healing."

explanation.⁶ Spirit possession is, most commonly, a subaltern discourse that communicates to the community of the caring (i.e., the congregation in this case) that a serious problem or set of problems has arisen that requires a significant amount of work to redress. In this way, public acknowledgment of one's possession is therapeutic in a performative sense by shifting one's illness to a public realm of discourse to be understood, managed, and positively affected by this community.⁷ Furthermore, spirit possession connotes misfortune or sickness, which consistently afflicts the marginalized in society. And in Africa, where patriarchy is a common feature of most societies, women are significantly marginalized and many become spirit-possessed.

Because affliction and healing are rooted in society, when society shifts through the process of migration, for instance, so do affliction and healing.⁸ But what aspect (or aspects) of society has changed during this particular migration to affect this particular healing institution? I argue that the gender-related transformation within this Christian healing institution is reflective of wider shifts in gendered power relations within Ghanaian social relationships, particularly, but not exclusively, at the level of conjugal relationships.⁹ In the United States, both social relationships and healing rituals within the Ghanaian Presbyterian community are marked by an increase in gender egalitarianism.

The Production of Patriarchy and Female Marginalization in Ghana

While precolonial Akan society (like today) was socially organized along matrilineal lines, it was not matriarchal. Women's structural centrality in relationship to descent and social organization was not

---

6. See Boddy, *Wombs and Alien Spirits*.
7. Ibid., 156.
8. Feierman and Janzen, "Introduction," 5.
9. Bettina Schmidt predicted this sort of transformation—differences in the gender distribution of spirit possession—as various cultural and sociopolitical factors change with respect to gender in the context of transnational migration ("Possessed Women," 112–13). Marja Tiilikainen has also recorded transformations of spirit possession practices among Somali immigrant women in Europe and North America ("Somali *Saar*").

reflected in their legal status.[10] Women were perpetual jural minors; they fell under the legal guardianship of a man regardless of age, individual wealth, or descent.[11] Freeborn men controlled the labor of women within their lineage. Furthermore, women's sexual rights were considered the property of their husbands. For example, the jural institution of *ayefere sika* was a compensatory fine for adultery paid by the adulterating male to the husband of the woman he had sex with.[12]

Many of these patriarchal relationships were maintained from the precolonial to the colonial period as the British strengthened non-capitalist social relations believed to be more conducive to the cheap production of agricultural commodities, particularly cocoa.[13] Ghana became the largest producer of cocoa in the world by 1911, and by 1920, cocoa amounted to 83 percent of Ghana's export earnings.[14] Nearly everyone in southern Ghana became involved in this lucrative industry. Land was being individually bought, sold, and mortgaged in the cocoa-growing forested districts of southern Ghana. Women played a central role in securing loans as pawns for male kin who needed capital to buy land for cocoa farms. Women also worked as porters, head-loading cocoa to market centers, as well as family laborers on cocoa farms. As jural minors, women were typically used in these capacities by senior male kinsmen—frequently by husbands—as an unpaid labor force. The exploitation of women's labor was greatly intensified during the colonial period, when women accounted for a significant amount of the productive forces within the cocoa economy.

The cocoa economy also increased women's hardships in at least four other ways besides contributing free labor.[15] First, cocoa farming was so labor intensive that women typically had less time to participate in income-producing activities of their own. Second, profits that

---

10. There were some areas of social life where relative gender equality existed in precolonial Akan society. Among the upper classes, women as well as men could occupy high political offices in old age (Akyeampong and Obeng, "Spirituality, Gender," 492). Among the peasant class in the subsistence-based economy, women and men produced agricultural crops on matrilineal land for personal or village consumption. The produce was shared. And married women during this time typically resided and labored on land in matrilineal villages with the support of their maternal kin.
11. Grier, "Pawns, Porters," 313.
12. Obeng, "Gendered Nationalism," 194.
13. Grier, "Pawns, Porters."
14. Kay, *Political Economy*, 15.
15. See Allman and Tashjian, "I Will Not Eat Stone."

husbands garnered were frequently directed toward the husband's matrilineage and not his conjugal family. Third, cocoa farming frequently involved migrating long distances from home villages, which removed women from the social support networks of female relatives, who helped with child rearing and maintaining the home. Fourth, support from husbands as well as brothers and maternal uncles—men within a woman's matrilineage—became much less predictable during the colonial period.

Many people who lived through the colonial cocoa boom remarked that "cocoa destroys kinship, and divides blood relations."[16] The capitalist cocoa economy weakened the matrilineage by extracting the primary work force of the lineage—young men and women—from lineage-controlled land, as well as compensating these workers with cash that was not subject to control by lineage elders. Cocoa and capitalism, therefore, strengthened the power of individual male farmers to the detriment of women and the matrilineage. The result was that southern Ghanaian society became much more patriarchal.

Many of these forms of patriarchy, a byproduct of the social changes within the colonial period, extended to the independence era and the present day. Opportunities for most women to earn relatively high levels of income are restricted because of low educational attainment and fewer opportunities for employment than men.[17] Much of the resources earned by a man are allocated to his matrilineage and are not invested in the conjugal family. As a result, women must rely on their maternal (and at times paternal) families for shelter and assistance in times of need. Unfortunately, however, access to inherited resources within the matrilineage are limited because of women's loss of inheritance rights over time, as preferences are given to male heirs, particularly sons who help to labor on their fathers' cocoa farms.[18] In many instances, therefore, women must subsidize the insufficient resources from their husbands with petty trade or agricultural work.

In cities and towns, women frequently work as petty traders, but men have recently made inroads into this informal economy, successfully competing with women in retail trade.[19] In rural areas, women are more likely to enter the cocoa industry today as farmers, but their

---

16. Ibid., 123; Mikell, *Cocoa and Chaos*, 112.
17. Manuh, "This Place," 82.
18. Mikell, *Cocoa and Chaos*, 112–23.
19. Overa, "When Men."

income is not likely to increase intrahousehold gender equality.[20] For the most part, female farmers have been excluded from the benefits of trade reform in the cocoa sector, even while obstacles to engage in cocoa production have been reduced for them.[21] And in these agricultural regions, when drought occurs, women are more vulnerable than men because of their combined reproductive and productive workload.[22] In Ghana, researchers have begun to refer to this situation as "the feminization of poverty."[23]

The state in Ghana also reinforces the subordination of women to men in society and interpersonal relations. Political life in Ghana is generally male dominated, and legislation reflects this.[24] While recent research has shown that one in three women in Ghana had experienced some form of violence within intimate relationships, recent efforts by foreign and domestic NGOs to pass a domestic violence bill, which criminalized domestic violence, including marital rape, failed because of government opposition.[25]

Christianity—the dominant religion in southern Ghana—has also played a role in the production of patriarchy. Women's power within Christian institutions is mediated primarily through their relationships to powerful men.[26] And while female prophets (ɔkɔmfo) have a considerable amount of religious authority within Akan religious traditions, this power has been generally usurped by the rapid spread of (patriarchal) Christianity in the twentieth century.

Female Spirit Possession and Male Deliverance Practitioners in Ghana

Capitalism, the state, and Christianity have contributed to the production of patriarchy in Ghana, which has marginalized women and produced a significant amount of social tension in their lives. Capitalistic

---

20. Ackah and Lay, "Gender Impacts," 218.
21. Ibid., 230.
22. Arku and Arku, "I Cannot Drink."
23. Wrigley-Asante, "Men Are Poor"; Yeboah, "Urban Poverty."
24. Manuah, "Doing Gender Work," 130.
25. Hodzic, "Unsettling Power." This case highlights, however, the growing number of feminist organizations that have emerged in Ghana during the last decade to address women's rights or to improve services for women (Hodzic, "Unsettling Power"; Manuh, "Doing Gender Work"; Tsikata, "Women's Organizing").
26. Soothill, *Gender, Social Change*, 138–39.

forms of production and domestic organization in particular have led the matrilineage to be viewed as morally ambivalent. On the one hand, capitalism (and Christianity) promotes the rupture of reciprocal obligations of the matrilineage by supporting individual wealth attainment to the detriment of the subsistence-based rural economy, of which the primary economic unit is the matrilineage. On the other hand, the hardships created by capitalism forces reciprocal aid within the matrilineage for survival.

Birgit Meyer argues that female spirit possession expresses the tension between the matrilineage and the individual in a capitalist economy in the context of Ghanaian Christian deliverance practices.[27] Many women attempt to sever kinship obligations through conversion to Christianity, which demonizes non-Christian rites that reinforce matrilineal ties.[28] However, the attempt to break socially or economically with the matrilineage is rarely successful, thereby producing a stressful situation in which many women become possessed by matrilineally associated spirits, such as family or clan deities.[29] Tensions between the matrilineage and the conjugal family, which are a result of and contribute to women's marginalization, are thus expressed through female spirit possession.

Meyer not only explains how and why women predominately become spirit-possessed, but also observes that men almost always manage women's possession among Ewe Christians in Ghana.[30]

---

27. Meyer, *Translating the Devil*. Although making little mention of gender, the colonial anthropologist R. S. Rattray noticed in the 1920s an association between marginalization and spirit possession in Ghana, inferring that the verb kɔm, meaning to be spirit-possessed, was etymologically related to the noun ɔkɔm, meaning to be hungry (*Religion and Art*, 42–43).

28. Engelke, "Discontinuity and the Discourse"; Meyer, "Make a Complete Break." In Ghana there are structural differences between indigenous healing and Christian therapeutics. Both forms of therapy agree that conflicted relationships between the individual and the lineage cause disease or misfortune. Leith Mullings (*Therapy, Ideology*) notes that indigenous therapies try to reconcile these conflicted relationships, whereas Christian therapy focused on severing these relationships. Members of the kinship group act as managers in indigenous therapy, whereas in Christian therapy the afflicted are made to abstain from contact with kin, while church leaders and members act as therapeutic managers.

29. Women are also often afflicted, but not possessed, by witches, who are believed to be primarily effective within one's matrilineage. This has led Peter Geschiere to argue that witchcraft in West Africa represents the "dark side of kinship" (*Modernity of Witchcraft*, 11).

30. Meyer, *Translating the Devil*, 160.

The division of gendered labor within deliverance practices among Ewe Christians in Ghana holds true for Akan Christians, particularly within Grace Presbyterian Church in Akropong. At Akropong, I witnessed over twenty-five spirit possession cases in the course of a month in February 2007. They took place during public deliverance services performed by Catechist Abboah-Offei, during private deliverance consultations at Grace Church, and during private consultations by people seeking help who came directly to Abboah-Offei's home. Every person possessed was a woman.

Conversely, all the pastors and deliverance practitioners—those who physically managed the possessed—were men. The men perform the ritualized aspects of the ministry. Some men act as deliverance counselors during private consultation, where the suffering meet with deliverance counselors individually. During public deliverance services, the primary deliverance practitioners—those praying for the afflicted, laying hands on them, and anointing them with oil—are almost always men. The secondary deliverance practitioners—who care for the spirit-possessed during these services by restraining them, lowering them to the floor, holding their babies before they are prayed for, cleaning up their vomit, and covering them with blankets while they experience possession—are also usually men. While many women do serve on Abboah-Offei's Deliverance Team and pray with other men as intercessors, the exclusive domain of women is among the spirit-possessed.

But among Ghanaian Christian immigrants in the United States, the situation is very different. All areas of social life have become more gender egalitarian: economically in the workplace, socially within the home, and religiously in the church. I will first discuss the changes in the gendered division of labor among Ghanaian immigrants within conjugal relationships both in the workplace and within the home.

Decreased Patriarchy among Ghanaian Immigrants in the Workplace and Home

The recent immigration of Ghanaians to the United States has produced a very different division of gendered labor that is characterized by a relative gender egalitarianism reflected in the host society.[31] The primary reason for this change has been the economic advancement

---

31. Between the 1970s and the 1990s, Americans became more comfortable with women, and in particular women with children, working at least part time when

of Ghanaian women working in the healthcare industry generally and nursing particularly.[32] This part of the "global care chain" represents a major global labor circuit within an emerging international division of labor.[33] As part of the new serving class in the global North, Ghanaian immigrant women have emerged as the systemic equivalent of an offshore proletariat.[34]

Forces both external and internal to Ghana have produced this economic circumstance. Externally, the United States is currently experiencing a high demand for skilled work in traditionally feminized sectors such as healthcare, where positions are increasingly filled by skilled female immigrants.[35] In particular, severe shortages in the US nursing workforce—estimated to be 1.2 million by 2014[36]—have prompted the importation of many immigrant nurses via private recruiting firms, enabled by US immigration legislation.[37] By 2000, the United States became the world's largest importer of nurses, over 90 percent of whom were women and 80 percent were from the global South.[38] While the largest number come from the Philippines, trained Ghanaian nurses have also taken advantage of these opportunities in the United States. Many other Ghanaians obtain nursing training once in the United States.

In Ghana, Structural Adjustment Programs instituted in 1983 significantly decreased national expenditure on healthcare, which has created large numbers of unemployed health professionals. This in turn has created a push toward emigration of health professionals. In 2000, over five hundred nurses left Ghana for employment abroad, which was more than twice the number of nurses graduating from Ghanaian universities that year.[39] This situation is not unique

---

their children were young, as well as men sharing household duties (Bolzendahl and Myers, "Feminist Attitudes").

32. As of 2007, 28 percent of all African immigrant women worked in the healthcare industry in the United States (Terrazas, "African Immigrants").

33. Hochschild, "Love and Gold," 17.

34. Sassen, "Global Migration."

35. Kofman, "Gendered Global Migrations."

36. Hecker, "Occupational Employment Projections."

37. The Immigration Nursing Relief Act of 1989 allowed for the importation of foreign nurses. Also, a provision was included on a draft of a US immigration bill (S2611) that would remove the cap on special visas for foreign nurses, allowing open entry to all qualified foreign nurses (Kingma, "Nurses on the Move").

38. Aiken, "U.S. Nurse Labor," 1299, 1308.

39. Kingma, "Nurses on the Move."

to Ghana: an estimated thirty thousand sub-Saharan African nurses and midwives educated in their home countries are now employed in Europe and North America, particularly in the United States.[40]

The predominance of Ghanaian Presbyterian women within the healthcare industry in North America was supported by the survey data I collected in November 2006 at the New York deliverance workshop, attended by various members of all the Ghanaian Presbyterian congregations in the United States and Canada.[41] Twenty of the thirty female respondents worked in some capacity within the healthcare industry. Of these twenty women, nine were nurses, nurse's assistants, or nursing students. In contrast, the employment trend among Ghanaian men was to work in the transportation or machinery industries as cab drivers, truck technicians, and machine operators. These professions accounted for seven of eighteen male responses.[42] The average earning potential in the nursing industry—particularly that of registered nurses (RN) and licensed practical nurses (LPN)—is greater than that of various transportation/machinery professions, although some men did have higher paying jobs in areas such as accounting and insurance.[43] A few men also worked in the healthcare industry.[44]

---

40. Ibid., 1282. This emigration has spurred debates over the ethics of large-scale emigration of health professionals. Some scholars argue that the unmanaged outflow of nurses from countries experiencing a shortage of healthcare providers is inconsistent with "health for all" principles and impairs the ability of those states to deliver vital health services to local communities (McElmurry et al., "Ethical Concerns"; Dovlo, "Migration of Nurses"). Others, while recognizing these problems, point out that remittances from nurses working abroad provide significant benefits, and their opportunity to develop skills could improve healthcare in their home countries, assuming they return (Ross, Polsky, and Sochalski, "Nursing Shortages"; Pittman, Aiken, and Buchan, "International Migration").

41. See Regina Gemignani, "Gender, Identity," for a study of Nigerian Christian women in the US healthcare industry.

42. Other male responses included youth music director, actuary, security officer, pastor, pastry chef, and accountant.

43. The average yearly salary in Philadelphia for nursing ranges from $27,000 for a nursing assistant and $40,000 for a licensed practical nurse to $63,000 for a registered nurse. Conversely, the male-oriented jobs range from $32,000 for taxi and light truck drivers to $40,000 for heavy truck drivers. These figures were derived from using the website salary.com (http://www.salary.com/category/salary/) for these particular jobs in Philadelphia in 2010.

44. Four of the nineteen men worked in the healthcare industry in the following jobs: nurse, preoperative clinic coordinator, mental healthcare worker, and nurse's

In many of these Ghanaian Presbyterian households, the wife's income was greater than her husband's, and in the majority, the spouses' two incomes were nearly equivalent. This relative gender egalitarianism has led to new problems within Ghanaian conjugal relationships in the diaspora. As Ghanaian wives work the same hours and earn equal or more money than their husbands, many Ghanaian men are expected to participate in the domestic sphere. Most are not eager to do so. Disagreements revolve around creating and maintaining joint bank accounts and deciding who gets to spend the money and how it should be spent.[45] Takyiwaa Manuh noted that many Ghanaian women in Toronto complained that their husbands tended to treat joint accounts as their own, making withdrawals without their wives' consent.[46] Their husbands would then use the money for their own benefit or the benefit of their extended families in Ghana. Conflict between the needs of the conjugal family and those of extended families on both sides are frequent areas of dispute.[47] Many Ghanaian women in the United States have much more power within these areas of conflict because of their financial power.

Female economic empowerment has been accompanied by female sociopolitical empowerment within Ghanaian conjugal relationships. This shift in sociopolitical power toward gender egalitarianism within Ghanaian homes in the United States has fostered discussions about the proper way for husbands and wives to interact in America, particularly in condemning spousal abuse within the Ghanaian Presbyterian community. For instance, during a workshop on marriage given to this community in Philadelphia, Rev. Margaret Asabea Aboagye, pastor of the Ghanaian Presbyterian Church in Columbus, Ohio, publicly denounced spousal abuse during a formal presentation, pointing toward worldly (the state)

---

assistant. Many Ghanaian men are now receiving training in the United States to enter the healthcare industry.

45. Some men contended that because they immigrated first and later brought their wives over, they were entitled to control their wages and demanded that their wives hand over their checks to them (Manuh, "This Place Is Not Ghana," 87; "Ghanaian Migrants," 103).

46. Manuh, "This Place Is Not Ghana," 87.

47. These divisions are particularly evident upon the death of elderly relatives in Ghana, when adult children in the United States are primarily responsible for funding expensive funerals for parents and other family members. Negotiating amounts to spend on a deceased parent's funeral in Ghana is a frequent area of dispute within conjugal relationships in the United States.

and otherworldly (God) punishment for abusive husbands. Toward the end of her presentation Aboagye stated, "I heard some men beat their wives [in Philadelphia]. Shame on you! If you beat your spouse, shame on you! It is not only the police, but also God should break your neck! In American marriages, the two have become one. How can you slap your own face?"[48] Aboagye argued that gender egalitarianism—where spousal abuse is absent—must exist within Ghanaian marriages because both husbands and wives have to rely on one another to survive in the United States. Voices of female authority within Ghanaian Christianity—demanding gender equality—are very uncommon in Ghana.

Furthermore, police intervention of domestic violence, publically discussed by Aboagye in the above presentation, is part of a larger discussion in which, in the United States as opposed to Ghana, the state is perceived to support women to the detriment of men. In particular, the state provides social services for victims of domestic violence, which are almost always women. For example, in April 2010 the Family Violence Option was officially codified in Pennsylvania, ensuring that the Commonwealth of Pennsylvania provides critical support and protection to domestic violence victims who need public assistance in order to help them achieve economic independence.[49] Not only are abused wives given assistance if they leave the home, but many in the Ghanaian community have also commented that when the police are called to intervene during episodes of spousal abuse, it is almost always men that are forced to leave the home. Therefore, in areas of most direct and physical conflict between husbands and wives, the state is believed to support women in the United States.

Gender, Power, and Healing in Ghanaian Presbyterian Churches in the United States

Immigration frequently alters gender relationships, most commonly by enhancing women's status owing to immigrant women's ease of entry into the US labor market. This in turn raises women's status relative to that in their home countries, while subsequently

---

48. Lecture on marriage given by Margaret Asabea Aboagye, Philadelphia, August 11, 2007.
49. Women's Law Project, "Violence Against Women."

decreasing men's status.[50] This shift in power frequently extends to the religious sphere in the context of immigrant congregations, where women's religious roles have grown even within strongly patriarchal groups.[51] This change in social, economic, and religious power between men and women holds true for the Ghanaian Presbyterians in the United States.[52]

In the three subsections that follow, I will describe areas of gendered divergence within religious practices of the Ghanaian Presbyterian community in the United States. First, I will discuss the increased leadership role of women as pastors. Second, I will describe how women have taken a greater role as deliverance practitioners. Third, I will explain the increased incidence of spirit possession among men.

Ghanaian Women as Pastors (and Healthcare Professionals)

While very few women are Presbyterian pastors in Ghana, by 2009 there were three female pastors within the Ghanaian Presbyterian community in the United States. I will discuss two of them in this subsection.[53] Rev. Rosamund Atta-Fynn is an ordained pastor and deacon in the United Ghanaian Community Church (UGCC), the Ghanaian Presbyterian Church in Philadelphia. Rev. Margaret Asabea Aboagye shares pastoral responsibility at the Ramseyer Presbyterian Church in Columbus, Ohio, with her husband, Rev. Andrew Aboagye. The two women share a few similarities: both are ordained pastors like their husbands, both work in the healthcare industry but have never introduced deliverance practices into their workplaces, and both are very active in their respective church's prayer team.

In Ghana, Atta-Fynn never thought of herself as someone who could become a pastor.[54] She saw herself as a teacher, which was her

---

50. Babou, "Migration and Cultural Change"; Mahler, "Engendering Transnational Migration"; Pessar, "On the Homefront"; "Engendering Migration Studies"; Zentgraf, "Immigration and Women's Empowerment." This claim assumes, however, that immigrants are coming from states considered more patriarchal than the United States, unlike some northern European countries, like Sweden, that are considered more gender egalitarian than the United States.

51. Boddy, "Managing Trandition"; Kim and Kim, "Ethnic Role."

52. See Biney, *From Africa to America*.

53. The third is Rev. Elizabeth A. Acquah, who leads the Presbyterian Church of Ghana Fellowship in Atlanta.

54. Author interview with Rosamund Atta-Fynn, Philadelphia, August 31, 2006.

profession in Ghana, as well as a pastor's wife. But when she immigrated to the United States, Atta-Fynn was encouraged to complete a seminary degree after taking a few courses at Eastern Baptist Theological Seminary (now called Palmer Theological Seminary) in suburban Philadelphia, where her husband had been a student. At Eastern Baptist, Atta-Fynn majored in theological studies and counseling, which led her to become a drug and alcohol rehabilitation counselor for the Philadelphia Mental Health Consortium. Beyond her regular duties, Atta-Fynn leads a special group session for Christians on Fridays based on the Alcoholics Anonymous twelve-step program. In these sessions she has not introduced deliverance practices, although she is cognizant of the twelve-step program's origins as a form of Christian therapy.[55] Atta-Fynn suggests that her Friday session is a more authentic version of the original program than the de-Christianized version more typically practiced in the United States.

Aboagye, as opposed to Atta-Fynn, did her seminary training in Ghana at Trinity Theological Seminary at the University of Ghana (where Ofosu-Donkoh was also trained) before emigrating to the United States. At that time in the 1980s, there were very few woman ordained in the Presbyterian Church—fewer than five, according to Aboagye. After working in certain female-specific roles in the Presbyterian Church of Ghana, such as chaplain at Aburi Women's Training College, she received a fellowship to study at McCormick Theological Seminary in Chicago, where she became particularly drawn to the hospital ministry. Soon after her arrival in the United States, she began dating Rev. Andrew Aboagye, who was leading the Ghanaian Presbyterian Church in the Bronx, New York. Together, they decided to move to Columbus, Ohio. In Columbus, Margaret Aboagye began working as a hospital chaplain in the critical care unit of a local hospital, while her husband became the fulltime pastor of the Ebenezer Presbyterian Church in Columbus, a Ghanaian Presbyterian church that the Aboagyes founded in 2003–4. Like Atta-Fynn, Aboagye does not act as a deliverance practitioner in the hospital. Aboagye did, however, successfully revive a patient, who was in a coma and experiencing organ failure, through prayer.[56]

---

55. The twelve-step Alcoholics Anonymous program derived much of its Christian therapeutic philosophy from an evangelical Christian organization, the Oxford Group.

56. A woman was in the critical care unit after being on life support for five days, and her daughters were saying their goodbyes before taking their mother off life support. One of the daughters asked Rev. Aboagye to pray for her mother. Aboagye touched the comatose woman's right foot and began to pray for life to pass through

Atta-Fynn and Aboagye are two Ghanaian Presbyterian women who have become pastors in and leaders of the emerging network of Ghanaian Presbyterian churches in North America. While they rarely bring deliverance practices to bear at their workplaces in the healthcare industry—usually because of laws separating church and state—they are both very active in the prayer teams of their respective churches. In the following subsection, I will describe the role of Ghanaian women as deliverance practitioners in the United States.

Ghanaian Women as Deliverance Practitioners

In Ghana, a division of gendered labor exists at Grace Presbyterian Church in Akropong, where the deliverance practitioners are men and nearly every spirit-possessed person is a woman. The gendered division of labor within deliverance practices in the United States is very different.[57] For instance, in Philadelphia, women have been leaders in the UGCC's prayer team from its inception to its transition into a group that performs deliverance practices. In the UGCC, three women initiated the prayer team in its inception in 2000. Between 2005 and 2008, when the prayer team began incorporating deliverance practices taught by Abboah-Offei, two-thirds of the prayer-team members were women. And after Abboah-Offei's second evangelical campaign to Philadelphia in April 2006, it was Rosamund Atta-Fynn along with two men, Isaac Baah and Emmanuel Oppong-Agyare, who

---

her. Immediately, the woman's right leg began shaking. Aboagye then touched the woman's left foot and began to pray for her full recovery. Soon after, the woman's left leg started shaking. This happened on a Thursday. On Friday morning, when Aboagye was eating her breakfast at the hospital cafeteria, the daughters approached Aboagye and told her that their mother had made a full recovery, was conscious, and even walking. While Aboagye did not deliver this woman from demonic spirits through prayer, she did believe that the woman's physical problems were caused or aggravated by strained relationships: between the sick woman and God as well as hatred within the sick woman's family. Author interview with Margaret Asabea Aboagye, New York, November 30, 2006.

57. While the prayer teams in the United States are trained by Abboah-Offei, most members have never been to Grace Church in Akropong. Subsequently, they are not aware of the division of gendered labor of the sixty-plus Grace Deliverance Team, nor does Abbaoh-Offei teach the importance of maintaining specific gender roles within the prayer teams in the United States.

began formally working as deliverance practitioners during public deliverance services and private individual consultations.

Ghanaian Presbyterian women in the United States, such as Rosamund Atta-Fynn and Margaret Asabea Aboagye, have begun participating in the ritualized aspects of deliverance practices, praying for the afflicted and caring for them if they become spirit-possessed. For example, in late December 2007, Catechist Abboah-Offei led a three-day revival service at the UGCC in Philadelphia. The first hour of this service on December 21 was allotted to singing and prayer, while the second hour consisted of a sermon about demonology: how spiritual exposure to non-Christian deities can have negative consequences in the lives of Christians. In the third hour, the theology of demonology was put into practice when a prayer line was formed; the afflicted and possessed congregants formed three lines, each consisting of twelve to fifteen people, down the center aisle of the sanctuary facing the alter. Facing the three lines of suffering Christians were three deliverance practitioners with their backs against the alter: Mr. Samuel Asare, who is Abboah-Offei's assistant at Grace Church; Dr. Rev. Kobina Ofosu-Donkoh, who is pastor of the UGCC; and Rev. Rosamund Atta-Fynn.

Atta-Fynn prayed with and for the afflicted individuals one by one, laying hands on them and anointing some with oil (see figures 6.1). Standing behind the suffering congregants were several secondary deliverance practitioners, who managed and cared for the afflicted when they became possessed. These included Isaac Baah, and also Ama Brown and Gloria Baden, two female prayer-team members. These two women laid hands on the afflicted, prayed for them, helped restrain the possessed, gently lowered the possessed to the floor if consciousness was lost, and cleaned up any fluids expunged during possession. One of the most significant transformations in the establishment of Christian healing practices among Ghanaian Presbyterians in the United States has been the significant increase of female deliverance practitioners.

Male Spirit Possession among the Ghanaian Presbyterian Diaspora

Gender transformations in the diaspora have extended from practitioner to patient; research in Philadelphia from 2005 through 2009, as well as in New York from 2006 through 2008, indicates that

Figure 6.1. Public deliverance practitioner Rev. Rosamund Atta-Fynn (standing on the right) at the United Ghanaian Community Church on December 21, 2007. Photograph by Kwame Ofosu-Donkoh. Reproduced with permission.

approximately 25 percent of all possession cases in the Ghanaian Presbyterian diaspora were men. Many of these men feel emasculated and marginalized by the economic independence of their wives and by their participation in domestic duties such as childrearing, food preparation, and cleaning. This decrease in their authority within their marriages, combined with increased expectations by extended family for greater financial support from their higher-paying overseas jobs, have led many Ghanaian men in the Presbyterian community to exhibit a traditionally feminine religious response to marginalization, namely, spirit possession.[58] While men get possessed less frequently and less violently than women in the United States, they are possessed far more frequently than in Ghana.

---

58. Men account for 58 percent of Ghanaian remitters, as opposed to Ghanaian women, who comprise the remaining 42 percent (Mazzucato, van den Boom, and Nsowah-Nuamah, "Origin and Destination," 141).

One example occurred at the UGCC, during a week-long healing revival held by Abboah-Offei in April 2006. A member of the church, David Frimpong (pseudonym), had been experiencing chronic illness and misfortune since immigrating to the United States in 1996. Frimpong had frequent stomach pains and was once taken to the emergency room, where doctors were unable diagnose his acute pain. His first child was born prematurely and almost died. Frimpong continually performed poorly in school and at work owing to poor concentration. In the course of one year, he bought six different cars and each one broke down. His marriage was under duress; Frimpong and his wife frequently argued over managing their finances and dividing household duties.

Eventually, Frimpong began asking his family in Ghana about possible causes of his illness and misfortune. Frimpong's mother told him that his disorder was spiritual and originated from within his extended family. Frimpong's sister-in-law became jealous that her husband had financially facilitated Frimpong's immigration to the United States. As a result of her jealousy, she attempted to harm or kill Frimpong by acquiring the services of an *aduruyefo*, an Akan spiritual practitioner known to harm people with spiritual medicine and a close approximation to a "sorcerer" in the ethnographic literature of the African occult.[59] Because Frimpong's afflictions were spiritual, he came for deliverance during a week-long healing revival held at the Philadelphia church hosted by Catechist Abboah-Offei.

Frimpong met with Abboah-Offei briefly for an individual consultation during the morning of April 25, 2006. Frimpong filled out a deliverance questionnaire and then discussed his situation briefly with Abboah-Offei. The catechist said that Frimpong was surrounded by a dark cloud, which was binding him. Abboah-Offei then prayed with him, after which Frimpong became possessed and lost consciousness. When Frimpong regained consciousness, Abboah-Offei anointed him with olive oil and suggested that he return to the deliverance service that evening.

---

59. Evans-Pritchard, "Sorcery and Native Opinion." Etymologically, the term *aduruyefo* is derived from *aduru*, meaning medicine, drug, powder, or poison, and *ayefo*, which is a maker, author, mischief-maker, or mischievous enemy (Christaller, *Dictionary*, 101, 587). This maker of medicine harnessed an ambivalent power that could be used to either cure or inflict harm. This general definition and etymology applies today.

During the middle of the evening service, Abboah-Offei led a prayer in which he called for the spirit of failure to be broken in the name of Jesus. The catechist signaled to two prayer-team members and Samuel Asare to usher Frimpong from a pew toward the rear of the sanctuary into the central isle. Abboah-Offei began to shout, "Break, break, break that darkness hovering around you!" Frimpong's face became contorted as he slowly fell backward, throwing his hands in the air and shouting. Frimpong became possessed, lost consciousness, and was lowered to the floor by the prayer-team members. He lay still on the floor for about ten minutes while being tended to by the two prayer-team members. Slowly he moved back to his seat, but remained visibly dazed for the rest of the service.

A similar set of events occurred at a service four days later on April 29 when Frimpong returned for the evening service. He recalled praying in the audience, which was followed by the experience of something moving inside him that held him down and burned. This burning sensation was particularly strong in his stomach. He could barely stand upright and held onto the pew tightly. Eventually, Frimpong stumbled into the aisle and fell down possessed, flailing his arms and foaming at the mouth. Abboah-Offei approached Frimpong, who was lying on the ground, laid hands on him, and ordered the demonic bondage emanating from the *aduruyefo* to be broken in the name of Jesus. In Frimpong's own words, Abboah-Offei breathed the Holy Spirit into him, which expelled the demons.[60] Examples of male spirit possession such as Frimpong's rarely, if ever, occur within the Presbyterian Church of Ghana or other Christian denominations in Ghana.

After this event, Frimpong began coming to Wednesday prayer services, and prayer-team members, particularly Isaac Baah, Rosamund Atta-Fynn, and Emmanuel Oppong-Agyare, began praying with him. After this deliverance event and continuous prayer for several weeks, Frimpong was finally freed from the power of the *aduruyefo*. By the following summer, Frimpong became the ninth member of the Philadelphia prayer team and on January 14, 2007, David Frimpong was installed as a deacon in the UGCC (see fig. 6.2).

---

60. Author interview with David Frimpong, Philadelphia, April 8, 2007. Abboah-Offei literally took large breaths of air and expelled them into Frimpong's mouth.

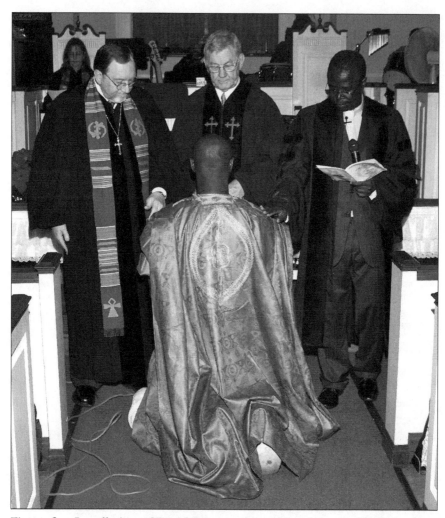

Figure 6.2. Installation of David Frimpong as a deacon in the United Ghanaian Community Church on January 14, 2007. *From left to right, standing:* Rev. Edward D. Gehres Jr., Rev. William Thompson, and Dr. Rev. Kobina Ofosu-Donkoh. Photograph by Kwame Ofosu-Donkoh. Reproduced with permission.

Conclusion: The Production of "Male Junior Femininity"

The significant amount of gender egalitarianism—social, economic, and religious—among Ghanaian Presbyterians in the United States is recognized and discussed within this community, not as gender egalitarianism per se, but as gender inversion. It is common to hear among Ghanaian Presbyterians that as a result of their immigration to the United States, men have become women and women have become men.[61] This transformation has been significant with regard to Ghanaian women, but radical with respect to Ghanaian men.

Ghanaian women have become socially empowered within their conjugal relationships as a result of their economic success in the United States. Ghanaian women in the United States, therefore, have attained a level of social and economic power similar to that of men in Ghana. A common Twi proverb that refers to similar shifts in power and corresponding gender ideology is *Obaa nya sika a, odane barima* (when a woman becomes wealthy, she turns into a man).[62] The existence of the proverb is an index that this transformation of Ghanaian women's power in the United States has existed in Ghana under certain conditions in the past. For instance, Pashington Obeng has described politically empowered women as embodying the gender category "female senior masculinity," which was achieved by some senior postmenopausal women who occupied male political offices in precolonial Ghana.[63]

Religiously, women have frequently served as Akan healers in traditional healing cults and shrines. Among the Christian community in Ghana, there are some instances in which women act as deliverance practitioners. For example, the largest church in Ghana, the Church of Pentecost, has a prayer camp (or healing center) in Cape Coast called Edumfa, led by a woman, Prophetess Boateng. Female healers or deliverance practitioners are not prominent, however, within the Presbyterian Church of Ghana.

Within the Ghanaian Presbyterian churches in the United States, the large percentage of female pastors and deliverance practitioners makes it significantly different from the Presbyterian Church in

---

61. Takyiwaa Manuh found the identical discourse among Ghanaian immigrants (not Presbyterians or Christians specifically) in Toronto ("This Place Is Not Ghana," 85).

62. Obeng, "Gendered Nationalism," 195.

63. Akyeampong and Obeng, "Spirituality, Gender"; Obeng, "Gendered Nationalism," 196–201. One of the primary case studies discussed by Pashington Obeng is Akyaawa Yikwan, who successfully negotiated a treaty with the British for the *Asantehene* (King of Asante) in 1831 (Wilks, *Forests of Gold*, 329–61). The Ghanaian historian and Basel Mission pastor Carl Reindorf wrote in 1895 that Akyaawa was a "woman of masculine spirit" (Reindorf, *History of the Gold Coast*, 198).

Ghana. In Ghana, a distinct gendered division of labor exists within the Presbyterian Church's deliverance ministry. Nearly all pastors, public deliverance practitioners, and deliverance counselors are men, while the great majority of the suffering who become spirit-possessed are women. In the United States, alternatively, this gendered division of labor has become significantly reduced. Women in the Ghanaian Presbyterian diaspora, such as Margaret Asabea Aboagye and Rosamund Atta-Fynn, have become ordained pastors as well as public deliverance practitioners and deliverance counselors. The wealth and social power accrued within conjugal relationships by Ghanaian Presbyterian women—combined with increased religious authority as pastors and deliverance practitioners—has resulted in a fairly widespread level of "female senior masculinity" within this community today.

As women have become healers, men have become beneficiaries of their ministrations; a significant minority of Ghanaian men have become spirit-possessed, a palpable sign of their marginalization in the United States. Spirit-possessed men are almost never found in Ghana, making this transformation significantly more radical than the occurrence of empowered women or female deliverance practitioners. There are three major differences between male spirit possession among the Ghanaian Presbyterian diaspora and previous research on male spirit possession among non-Christian religious groups in Africa and the African diaspora. One, the quantity of male spirit possession among the Ghanaian Presbyterian diaspora is significantly larger. Researchers such as I. M. Lewis have pointed to a small minority of men who were spirit-possessed. These men were subordinated in their society, such as Muslims or ex-slaves within the *Zar* cult in Christian Ethiopia, domestic servants or unemployed laborers within the *Shango* cult in Trinidad, or marginalized labor migrants within the *Bori* cult in Niger.[64] These spirit-possessed men were marginalized, but constituted a very small minority of the spirit-possessed, whereas among the Ghanaian Presbyterian diaspora spirit-possessed men constitute about 25 percent of all occurrences.

A second major difference is that spirit-possessed men of the Ghanaian Presbyterian diaspora are part of the religious orthodoxy. Spirit-possessed Muslim men taking part in the *Bori* spirit possession

---

64. Lewis, *Ecstatic Religion*, 100–107. See also, but to a lesser extent, Boddy, *Wombs and Alien Spirits*, 209–12; and Igreja, Dias-Lambranca, and Richters, "*Gamba* Spirits."

cult in northern Nigeria violate orthodox Islamic norms. Alternatively, among Pentecostals, Neo-Pentecostals, and Charismatic mainline churches like this Ghanaian Presbyterian community, spirit possession by demons and its treatment via deliverance practices is considered orthodox behavior. Spirit possession among men does not violate Christian (and particularly Pentecostal) orthodoxy.

A third major difference is that within the Ghanaian Presbyterian diaspora, there is no association between spirit-possessed men and homosexuality or transvestism. The *'yan daudu* (or feminine men) of northern Nigeria, who are considered sexual deviants (pimps, prostitutes, and homosexuals) by mainstream society, are strongly associated with the *Bori* spirit possession cult.[65] In Brazilian *Candomblé*, to give another example, there is also a strong association (made externally by mainstream society) between possession priests, transvestism, and homosexuality, or more specifically men who get "mounted" (*adés*) by other men as well as spirits.[66] Alternatively, spirit-possessed men of the Ghanaian Presbyterian diaspora are never associated with homosexuality or transvestism.

While there is no association with homosexuality or transvestism—common signs of femininity—within the Ghanaian Presbyterian community, spirit possession has still retained its metacultural significance as a female-associated behavior, where spirit-possessed men are perceived as "acting like women." Signifying oneself as masculine—the dominant gender category—requires demonstrating that one possesses the capacities to, among other things, resist being dominated by others, including other spirits.[67] Within Ghanaian society, one important signifying manhood act—which constitutes the production of the social category of a man—is the absence or avoidance of spirit possession.[68]

Therefore, spirit-possessed men in the Ghanaian Presbyterian diaspora, I argue, embody the gender category "male junior femininity," being the inverse of "female senior masculinity." Let me briefly discuss the significance of junior status in this situation. In the United States, not every man in the Ghanaian Presbyterian community suffers equally; their marginalization is variable. On average,

---

65. Gaudio, *Allah Made Us*.
66. Matory, *Black Atlantic Religion*, 207–12.
67. Schrock and Schwalbe, "Men, Masculinity," 280.
68. Schwalbe, "Identity Stakes."

spirit-possessed men are younger, have more recently immigrated, and, like David Frimpong, are experiencing conflict within both their extended and conjugal families. So the phenomenon of male spirit possession also has a temporal aspect to it—not age specifically, but seniority in their new context—which is continuous with the ways in which gender roles have shifted with respect to age and seniority in Ghana.[69] Therefore, male spirit possession can be seen as a form of "male femininity" experienced specifically by juniors in the Ghanaian Presbyterian diaspora, whose marginalization is a palpable sign of their junior status. What J. Lorand Matory argues for Brazilian *Candomblé* holds true for Ghanaian Presbyterians; nothing conveys junior and subordinate status more than the tendency for a man to get possessed.[70]

---

69. See Miescher, "Becoming ɔpinyin"; Obeng, "Gendered Nationalism."
70. Matory, *Black Atlantic Religion*, 215.

Conclusion

In this book I have argued that the lives of Ghanaian Presbyterians have become significantly more enchanted the more they have become incorporated into capitalist modes of production, particularly in the context of labor migration, over both time and space. This finding is ironic in light of Max Weber's most famous argument: that early Calvinist communities, of which Presbyterianism is a type, gave rise to the particular form of modern capitalism, but this economic system in turn destroyed the religious foundations that led to its emergence. Weber terms this form of religious destruction "disenchantment."

Weber has a very nuanced meaning of religious disenchantment, which refers to three interrelated concepts. One, disenchantment signifies the decline of supernatural modes of explanation from the world and their replacement with worldly explanations. Two, disenchantment refers to the departure of supernatural forces behind the natural world that interact in human affairs, such as spirits, demons, and gods. Three, disenchantment denotes the absence of charismatically endowed humans—referred to alternatively as sorcerers, magicians, and spiritual advisors—to manipulate these supernatural beings in accordance with human agency.

More specifically, however, I have applied Weber's understanding of religious enchantment to conceptions of illness, health, and healing. Therefore, as discussed in this book, enchantment means three things: the increase of supernatural explanations of illness, health, and healing; the intensification of spirits interacting with humans that alternatively afflict or heal; and the presence of charismatically endowed humans—referred to as deliverance practitioners—who manage human-spirit interactions for the betterment of afflicted individuals.

In southern Ghana, among the Akan, society was and continues to be religiously enchanted. What Kofi Appiah-Kubi wrote in 1981 holds true today; Akan society recognizes both biomedical and spiritual causes of illness and affliction.[1] Atheism is not a prevalent ontological position in Ghana, but is a specific Euro-American invention.[2] Historically, various churches that were established in Ghana have not always shared the same level of enchantment as mainstream Akan society. Most of the early mission churches established in the nineteenth century presented a fairly disenchanted form of Christianity in Ghana, particularly with respect to illness, health, and healing.

The Basel Mission was a notable exception. The Basel missionaries came from a region in southwest Germany that practiced a highly enchanted form of Christianity, exemplified by the healing ministry of Johann Christoph Blumhardt at his *Kurhaus* in Bad Boll. But these healing practices were never systematically replicated in Ghana; instead, both missionaries and Akan Christians consulted various Akan healers for relief of their suffering. After 1885, and significantly influenced by the medical discoveries of the bacteriological revolution, the Basel Mission began exclusively supporting biomedicine. Subsequently, the missionaries began excommunicating Ghanaian members who consulted Akan healers. At this point, after 1885, the Basel Mission (but not its congregants) in Ghana became institutionally disenchanted, and biomedicine, not capitalism, was the primary driving force.

By 1918, the tension between the enchanted Basel Christian community and the disenchanted Basel Mission in Ghana came to a head. In that year the Basel missionaries were expelled from the British colony, the deadly second wave of the 1918–19 influenza pandemic devastated Ghana, and cocoa production and prices on the world market plummeted, causing a short-term economic crisis that devastated many Basel Christians who were leaders in the burgeoning cocoa industry. These three acute crises were compounded by the social disruption produced by the cocoa industry and the correlating urban to rural labor migration, in which witchcraft accusations thrived. As a result, several enchanted churches were established and subsequently flourished in Ghana, most notably Faith Tabernacle Congregation, which systematically introduced Christian healing practices into Ghana. The confluence of these crises—which all revolved around colonial capitalism—led to the establishment of enchanted Christianity in Ghana.

---

1. Appiah-Kubi, *Man Cures*, 1–4.
2. Nancy, *Dis-Enclosure*, 14–15.

Faith Tabernacle's Pentecostal branches, particularly the Church of Pentecost, became so successful that by the 1960s the Presbyterian Church of Ghana's leadership introduced healing into the church. The main institution within the Presbyterian Church of Ghana established to provide healing to its congregants was the Bible Study and Prayer Group (BSPG). By the mid-1990s, Catechist Ebenezer Abboah-Offei, a prominent BSPG member, established the first charismatic Presbyterian Church in the town of Akropong, which served as a deliverance center: offering public deliverance services twice a week, individual consultation to the afflicted several times a week, and training deliverance practitioners from all over Ghana—and as far away as Nigeria—in healing and deliverance. By the turn of the twenty-first century, the Presbyterian Church of Ghana became significantly enchanted in the context of political, economic, and biomedical decline that created large-scale suffering for Ghanaians through the latter half of the twentieth century.

Not only did the Presbyterian Church of Ghana become enchanted by this time, but so did the Ghanaian Presbyterian churches in North America. These churches were established by Ghanaian labor migrants beginning in the 1980s. Corresponding with their migration, many in this community were experiencing spiritual afflictions, particularly by witches, that could only be addressed by spiritual means. By 2004, Catechist Abboah-Offei began offering a weeklong deliverance course to the prayer teams—the equivalent of the BSPG in North America—of these diasporic congregations. This training, to a great extent, established able deliverance practitioners within each of the Ghanaian Presbyterian churches in North America. The deliverance seminar or school of deliverance was the primary institution leading to the enchantment of the Ghanaian Presbyterian churches in North America.

In the Ghanaian Presbyterian Church in Philadelphia, called the United Ghanaian Community Church (UGCC), the prayer team was originally envisioned as one of the subchurch organizations by the founding pastor, Dr. Rev. Kobina Osofo-Donkoh. For nearly ten years after the establishment of the UGCC in 1995, the prayer team met periodically to pray for church members during Wednesday evenings. But as a result of two important week-long deliverance campaigns by Catechist Abboah-Offei in late 2005 and early 2006, the Philadelphia prayer team was transformed into a deliverance team that met with the suffering for individual consultations and periodically participated in public deliverance services. The institutionalization of deliverance practices in the UGCC, which formally enchanted this congregation, was further cemented when the entire prayer team was voted into the two governing bodies of the church in early 2007.

The replication of deliverance practices established in Ghana by Catechist Ebenezer Abboah-Offei helped to enchant the Ghanaian Presbyterian churches in the United States. However, there have been some notable disjunctures in this transnational process. In particular, gendered transformation of deliverance practices have been the most radical, where women in the Ghanaian Presbyterian diaspora have begun to work as deliverance practitioners and men, marginalized in the United States as a result of Ghanaian women's great economic success within the healthcare industry, have become spirit-possessed. Male spirit possession, an event almost never witnessed in Ghana, is an index of an emergent gender category in the diaspora that I call "male junior femininity."

In the two subsections that follow, I address two lingering questions in light of the developed proposition of this book. Why has biomedicine—one of Weber's two disenchanting forces along with capitalism—not disenchanted Akan society or the Ghanaian Presbyterian community particularly? What role does that state—through social welfare and land tenure policies—play in mediating this relationship between capitalism and enchantment?

Biomedicine's Negligible Effect on Religious Disenchantment

Throughout this book I have argued that the enchantment of the Presbyterian Church of Ghana, both in Ghana and in North America, has been engendered by the increase in capitalist modes of production, particularly under the conditions of labor migration. This relationship was found to be the inverse suggested by Max Weber nearly one hundred years ago, who argues that capitalism, emerging within the Puritan sects, was a disenchanting force in the world. But as I discussed in my introduction, Weber attributes two forces to disenchantment, capitalism *and* biomedicine. In chapter 1, I demonstrated the role that biomedical advances played in disenchanting the Basel Mission by significantly affecting the European missionaries' ideologies and practices of Christian healing. However, up to this point, I have said very little about how biomedicine possibly affected Ghanaian Presbyterians in Ghana or the diaspora. In the following subsection, I will explain how and why biomedicine did not become a disenchanting force among this transnational community.

In supporting this claim I will initially focus my attention on the local understandings of biomedicine in Ghana when this new technology was introduced during the colonial period, because many of these understandings of biomedicine persist today. I will also discuss particular biomedical issues that have come to fruition in the postcolonial period. At times during this discussion, I will vary the context between Ghana or Akan society and the Basel Mission/Presbyterian Church specifically, with the assumption that forces that affect Ghana or Akan society would also have an equivalent effect among Akan Christians of the Basel Mission/Presbyterian Church.

Biomedicine, like capitalism, has not disenchanted Akan society or Ghanaian Presbyterianism specifically. There are at least five reasons for this. One, biomedicine became associated with sorcery and, correspondingly, physicians became associated with sorcerers. Two, biomedicine was, at times, understood to be violent. Three, biomedicine, under certain conditions, was perceived to be exclusionary. Four, biomedical forms of economic transactions were believed to violate a local moral economy. And a fifth reason, which is more materially focused as opposed to cultural-critical, is that access to biomedicine was and continues to be very limited in Ghana.

Let me examine, first, how biomedicine became associated with sorcery. Biomedicine, when introduced to Ghana by medical missionaries and colonial physicians, became another variety of medicine available, which together was known as *aduru*. *Aduru*, however, was a very complex and diverse concept, which included powder (such as gunpowder), medicine, drugs, and particularly poison.[3] *Aduru* was an amoral, powerful force. It could be a benevolent medicine or an afflictive poison, and with it, practitioners of *aduru* could either heal or harm.

The original Twi translation of the word "physician," including instances in the Bible,[4] was *aduruyefo*, which most directly translates to "maker of medicine." The etymology of the term "medicine maker" comes from *aduru*, meaning medicine, drug, powder, or poison,

---

3. Christaller, *Dictionary*, 100–101.

4. For instance, Luke 8:43–48 describes a woman with a severe bleeding condition who had spent all her money on doctors (translated as *aduruyefo*) to no avail until she touched Jesus' cloak and was instantly healed.

and *ayefo*, which is a maker, author, mischief maker, or mischievous enemy.[5] The *aduruyefo* treated an array of personal and social afflictions such as sickness, war, wealth, pregnancy and birth, hunting and agriculture, success in business or at school, love affairs, and family conflicts.[6] But the *aduruyefo* among all Akan healers was particularly known to afflict its victims with spiritual poison. Subsequently, in the ethnographic literature, the word *aduruyefo* would translate to "sorcerer," and has been referred to in the David Frimpong case study in chapter 6. Thus, physicians and sorcerers were semantically linked as practitioners of potentially harmful or helpful medicine.

One particular example from Kwadjo Nti—the early Faith Tabernacle evangelist from Asienimpon (introduced in chapter 2)—demonstrates the complex understanding or potentialities of biomedicine in colonial Ghana.[7] In 1918, Nti had had a disagreement with the pastor of his Basel church during a meeting in Kumasi. Subsequently, the pastor cursed Nti three times. Immediately, Nti's health deteriorated and he was rushed to his home. From his home, Nti was brought to the Basel Mission hospital in Abetifi, where a European doctor saw him.[8] The physician promised to cure Nti, but after realizing he could not, made the decision to poison him. The physician mixed the poison in a cup for Nti to drink. Before drinking, however, Nti's nurse alerted him to the physician's devious plan and Nti secretly fled the hospital (and was later healed by prayer at Faith Tabernacle).[9] All the ambivalence of the *aduruyefo*—as it is understood locally—is embodied in this story about the European physician, who had the power to either cure or poison.

While accusations that physicians act as sorcerers are rare in this transnational Ghanaian Presbyterian community, some within the community suggest that physicians' abilities are limited and some even contend that the faithful should only rely on Christian prayer.

---

5. Christaller, *Dictionary*, 101, 587.

6. Rattray, *Religion and Art*, 40; Christaller, *Dictionary*, 101.

7. Okai, *General History*, 2–3.

8. By February 1918, the Basel Missionary staff, including the missionary doctors, were all expelled from colonial Ghana. This event, therefore, must have taken place sometime in the first several weeks of 1918.

9. What Franz Fanon writes about colonial Algeria holds true for colonial Ghana; a considerable amount of ambiguity existed between being medically treated (presumably helped) by one's colonizers, who were believed to be causing various types of political, economic, and social harm ("Medicine and Colonialism," 123).

One example came from a lecture given by John Gyadu (pseudonym) of the Ghanaian Presbyterian Church in Worcester, Massachusetts, during the 2006 deliverance workshop in New York. Gyadu testified to not using medicine for over two years; when sick he would put his hands together and pray. Furthermore, Gyadu described being significantly influenced in Christian biomedical abstinence by the writings of Smith Wigglesworth—a late nineteenth-century/early twentieth-century divine healing proponent—whose books Gyadu had purchased and were being circulated among others participating in the deliverance workshop. During this lecture, Gyadu recounted Luke 8:43–48, a passage in which a women with a chronic bleeding problem, after being unsuccessfully treated by a physician, was healed by Jesus after she touched his clothing. This account from Luke was employed to stress the enchanted nature of disease, the inability of physicians to address these issues, and the possible ways in which abstaining from biomedicine can help believers attain good health.

Two, biomedicine in Ghana has been understood to be violent under certain circumstances, besides its associations with sorcery and its limited effectiveness. The bacteriological revolution and the establishment of medical missions in Ghana corresponded with the formal colonial period in the 1880s, which disenfranchised most Africans. The biomedical improvements during this period—particularly the use of quinine by colonizers and missionaries as a prophylaxis against malaria—facilitated the political and economic advances of Europeans in sub-Saharan Africa as much as the Maxim gun. The colonial state's public health policies, such as sanitary segregation—meant to protect European populations in Africa from disease and particularly diseased Africans—had the unintended consequence of depriving African businessmen and businesswomen of equal competition with Europeans, which was also understood by Africans, at times, to be violent toward their economic well-being.[10]

Three, biomedicine carried with it a problematic excluding quality, which caused Ghanaian members of the Basel Mission a significant amount of distress. During its introduction, many Ghanaians were excluded from biomedical treatment; the medical missionary Rudolf Fisch's first order was to treat missionaries and other Europeans.[11] Once established, biomedicine as practiced within the Basel

---

10. Curtin, "Medical Knowledge," 243–49.
11. Fischer, *Der Missionsarzt*, 201.

Mission led to the exclusion of various forms of Akan therapy that many congregants relied on significantly for their health and welfare (see chapter 1). After 1885 and the arrival of Rudolf Fisch, the Basel Mission became disenchanted, which included denying the existence of many malevolent supernatural forces that afflicted Akan Christians, condemning Akan healers as illegitimate and anti-Christian, and excommunicating many Christians that visited Akan healers. To many Basel Christians, biomedicine was violent, exclusionary, and discriminatory.

Four, biomedicine in Ghana was problematic in that forms of biomedical payments violated a local moral economy akin to what Steven Feierman has demonstrated for Tanzania.[12] While most Akan healers accept a nominal payment to begin treatment, and ask for a more substantial payment when the patient is cured, the physician or hospital demands money up front whether healing occurs or not. Illness treated via biomedicine can financially ruin people and families. Many believe physicians to be uncaring, and also have moral problems paying for unsuccessful therapy.

A final reason explaining why biomedicine had a negligible effect on disenchanting Ghanaian society was and continues to be the lack of access to biomedicine in Ghana. During the colonial period, very few hospitals were established and few trained medical personnel were available, which limited local access to the benefits of biomedicine. This biomedical landscape has extended to the postcolonial period, in which access to biomedicine has remained a significant problem throughout Ghana; there are less than 1,440 healthcare facilities for a population just below 24 million people.[13] Furthermore, access to health services is geographically prohibitive, as most facilities are located in urban areas. Half the population of Ghana lives outside a 5 kilometer radius of a healthcare facility, which corresponds to a one hour walking distance, and a quarter of the population lives more than 15 kilometers from a facility in which a doctor can be consulted.[14]

Biomedical healthcare in Ghana is also financially prohibitive for many Ghanaians. The "cash-and-carry system" that existed from the 1980s into the first decade of the twenty-first century reinforced the

---

12. Feierman, "Explanation and Uncertainty," 341–44.
13. Austrian Red Cross, "Health Care in Ghana," 6–7.
14. Ibid.

problematic exclusionary relationship of the majority in Akan society to biomedicine. In particular, biomedicine has excluded many poor Ghanaians who could not afford to pay. Even with the 2003 introduction of a national health insurance scheme in Ghana, between 65 percent and 82 percent of the population remained uninsured by 2010. Many find the annual fee of around nine dollars prohibitive.[15] Some physicians have also denied care to insurance holders because of the long reimbursement delays by insurance providers.[16] While Ghanaians in the United States have access to quality healthcare with legal status and employment, many do not satisfy both requirements. And some illegal Ghanaian immigrants believe that visiting a hospital significantly increases one's risk of deportation (see case of Ama Asare in chapter 5).

For all the above reasons—semantically linking biomedicine with sorcery, politically and economically violent understandings of biomedicine, biomedicine's exclusionary practices, its violation of a local moral economy, and finally lack of access—biomedicine was never anchored in Ghanaian society as it was in the global North, nor did it have a disenchanting affect on society.

The Role of the State: Social Welfare, Land Tenure, and Religious Enchantment

I have so far argued that the Presbyterian Church of Ghana, in both Ghana and the United States, has become more enchanted the more its members have become integrated into capitalist modes of production. My proposition—like Weber's in *The Protestant Ethic and the Spirit of Capitalism*—is based on an explicit relationship between religion and economics. But how does politics, or the state specifically, fit into this equation? In this subsection, I will make two arguments about how the state affects the relationship between religious enchantment and capitalism. One, I claim that state welfare spending has an inverse relationship to religious enchantment; Ghana's low state welfare spending therefore effects its high level of religious enchantment. Two, I contend that the form religious enchantment takes in Ghana (and among Ghanaians living in the North America)

---

15. Kotin, "Health Insurance"; Oxfam "Don't Copy."
16. Kotin, "Health Insurance."

is affected by the Ghanaian state's corporate or customary land tenure system, which is a feature of most sub-Saharan African states.

Let me begin by discussing the relationship of welfare states to disenchantment. The welfare state is a historical product of Europe; Germany's Otto von Bismarck is credited with creating the first modern European welfare state. In the 1880s, Bismarck introduced old age pensions, accident insurance, medical care, and unemployment insurance, which became the model for other countries and the basis of the modern welfare state. One of Bismarck's goals in creating this type of social security net was to reduce the outflow of German immigrants to America, where wages were higher, but welfare did not exist.[17] Therefore, European welfare policies were created in reaction to, among other things, America's free market system that provided higher wages, but did not provide many social services.

The United States' lack of a strong socialist movement with corresponding welfare opportunities, compared to Europe, has been deemed "American exceptionalism."[18] More recently, Bryan Turner argues that "American exceptionalism" accounts for the exceptional difference between the role of religion in politics in American society compared with European society.[19] I believe these two concepts—welfare and religiosity—are integrally linked.

Anthony Gill and Erik Lundsgaarde have demonstrated that state welfare spending has a detrimental, although unintended, effect on long-term religious participation and overall religiosity.[20] This relationship accounts for why the United States is significantly more religious than Western European states, and also why Uruguay is significantly more secular[21] than its South American neighbors. America, unlike Western European states, spends relatively little on

---

17. Hennock, *Origins of the Welfare State*.
18. Lipset, *First New Nation*, vii, 178–79.
19. Turner, *Religion and Modern Society*, 141–46.
20. Gill and Lundsgaarde, "State Welfare Spending." In this study, three variables were used to measure religiosity: religious participation or attendance, the percentage of the population recorded as "nonreligious," and the proportion of the population that said they take comfort in religion. Alternatively, welfare was measured as the total government social welfare expenditures (including social security) divided by GDP and calculated on a per capita basis.
21. By "secularism," I mean a decline in church membership and attendance, the marginalization of the church from public life, and the dominance of scientific explanations—as opposed to spiritual explanations—of the world (Turner, *Religion and Modern Society*, 11).

welfare. In contrast to the United States, Uruguay has an extensive social welfare system. The positive relationship between increased welfare expenditure and secularism (as a specific type of religious disenchantment) is supported by a recent Gallop poll published in 2009 that asked a fairly general question—is religion important in your daily life?—to representatives from 143 countries and territories.[22] In Ghana, only 5 percent of those polled answered "no" to this question. Under conditions of structural adjustment, only 5.5 percent of Ghana's GDP was spent on primary education, health, and water and sanitation in 2004.[23] And in 2010, only 2.8 percent of Ghana's GDP was spent on healthcare.[24]

Historically, most Christian denominations have played a significant role in providing forms of social welfare, which has significantly affected religious involvement and belief. But under conditions of state-provided welfare—in which money is coerced through taxes (on threat of imprisonment) as opposed to voluntarily given through tithes and offerings—religious institutions have a difficult time competing with welfare states.[25] As in Northern Europe under conditions of robust welfare spending, citizens become quite secure in their day-to-day lives and religion loses much of its meaning and relevance. In acknowledgment of this relationship, Karl Marx proclaimed in 1844—before Bismarck's institution of social welfare in Germany—that "religion is the sigh of the oppressed creature, the sentiment of a heartless world, and the soul of soulless conditions."[26] The religious disenchantment lamented by Max Weber most frequently unfolds under conditions of capitalism and in the context of robust welfare spending.[27]

---

22. Crabtree and Pelham, "What Alabamians and Iranians Have in Common."
23. Bussolo and Medvedev, *Challenges to MDG.*
24. ISODEC-UNICEF, *Analysis*, 23.
25. Gill and Lundsgaarde, "State Welfare Spending," 408.
26. Marx, "Towards a Critique," 54.
27. Around the same time that Weber wrote about the disenchantment of the world at the turn of the twentieth century, Bismarck was instituting his welfare policies in recently unified Germany. Maybe, then, it was not individual wealth resulting from capitalism that was disenchanting German society at this time. Perhaps wealth was being redistributed via the state to German citizens, making them collectively more secure in this world in combination with accessible medical care. This medical care was particularly successful in light of the new medical technologies resulting from the bacteriological revolution, in no small part played by the microbiological discoveries of the German Robert Koch, who isolated the bacteria causing anthrax

In weaker welfare states or among marginalized citizens in stronger welfare states, religious institutions perform many of the same social security functions as robust welfare states. For instance, Philadelphia's UGCC provides several social security benefits to its members. When a church member dies, the church pays for the casket ($2,000) and takes a collection for funeral costs. If the deceased does not have relatives in the United States, the church pays all funeral expenses, including food, drink, local funeral expenses, and the cost of flying the body back to Ghana. If a member loses a conjugal family member in the United States or in Ghana, the church gives $500 to the family and collects donations from the congregation. If a member loses his or her job, the church takes collections for several weeks in order to provide money for the unemployed member. Sick individuals unable to work are awarded $100. The church gives an interest-free "soft loan" with flexible terms of payment to members in extreme cases. Women at the UGCC have also adopted the baby shower and bridal shower practices of American women, which provide them with additional forms of mutual aid and support. These benefits positively affect the overall religiosity of the congregation and are extremely important to these recent immigrants, some of whom are illegal and living under conditions of great insecurity. These social security measures are frequently interpreted as divinely given.

Not only is there an inverse relationship between state welfare spending and levels of enchantment in society, but the state also affects enchantment in another domain. The communal land tenure systems in Ghana play a qualitative role in the form that enchantment takes, because communal land tenure embeds social forms of obligation and reciprocity within the state. Forms of lineage-based obligation and reciprocity have cultural, social, and historical components, to be sure, but what I argue is that there is a political component with spiritual consequences; if these social forms of obligation and reciprocity are violated, as is frequently the case with Christian labor migrants, the result is often spiritual affliction.

In Africa today, communal or customary law continues to be the predominant form of governance, particularly with regard to land tenure, where property regimes are legitimated by customary kinship. Approximately 90 percent of land in Africa is owned communally

---

(1877), tuberculosis (1882), and cholera (1883) during the same time Bismarck was instituting these welfare policies.

and governed by customary law.²⁸ In Ghana specifically, 80 percent of all land is regulated by customary tenure controlled by traditional authorities (chiefs, priests, elders) in trust for decent groups.²⁹ And membership into a decent group or lineage (matrilineage in Ghana) guarantees any individual access to housing and typically some form of economic survival. As a result, social forms of obligation and reciprocity within the lineage are embedded within the state.

But capitalist labor migration disrupts this customary "economy of affection," destabilizing this kinship system.³⁰ In conjunction with Christianity, which stresses monogamy as well as distribution and care within the conjugal family, forms of capitalism weaken the lineage by pulling young men and women, the primary labor force on lineage-owned land, away from these traditional forms of community and obligation. Furthermore, money earned through capitalist endeavors on privately owned land is typically the property of an individual and not of a lineage.

This customary land tenure system and its obligations of reciprocity have led to resource battles within extended families.³¹ These conflicts are particularly acute among labor migrants, and Christian labor migrants more so, who are socially linked to their matrilineage but not fully participating in this resource-sharing system. And these resource struggles create the context for, and at times produce, spiritual afflictions, both in Ghana and among the Ghanaian Presbyterian diaspora. Malevolent spiritual forces that afflict Ghanaian Presbyterians, such as witches, are typically produced by living individuals within the sufferer's extended family.³² As Wim van Binsbergen has argued, the old kinship order and its associated forms of witchcraft accusations is never far away from the personal lives of the most urban—and I would argue transnational—Africans, regardless of class position.³³ Forms of witchcraft are intimately attached to land, ecology, and economic affairs even in the context of the transition to private property.³⁴ These afflictions are "socio-spiritual," a term that

---

28. Deininger, *Land Policy*.
29. Obeng-Odoom, "Land Reforms,"164; Ubink, "Negotiated or Negated," 266.
30. Hyden, *Beyond Ugamaa*.
31. Obeng-Odoom, "Land Reforms"; Ubink, "Negotiated or Negated."
32. This has been found to be a trend throughout sub-Saharan Africa (Westerlund, *African Indigenous Religion*).
33. Van Binsbergen, "Witchcraft in Modern Africa," 252.
34. Shipton, *Mortgaging the Ancestors*, 72.

refers to the dual dimension of these afflictions as spiritual entities that harm, which are also indicative or causative of social tensions, particularly within the sufferer's extended family.[35]

The socio-spiritual afflictions within the Ghanaian Presbyterian community described in this book demonstrate this complex issue of conflicted kinship, derived from social obligations embedded in the state apparatus. For example, Dr. Rev. Ofosu-Donkoh's maternal relatives bewitched and killed his mother after his birth. A man treated by Abboah-Offei during his 2005 evangelism trip to Philadelphia was inhibited from being successful in school because he had been cursed or bewitched by a family member in Ghana. Also, David Frimpong's series of misfortunes and afflictions were caused by an *aduruyefo* at the request of his jealous sister-in-law. Therefore, the world of enchanted Christian labor migration revolves around Jesus Christ, wage labor, distribution within their conjugal families and congregations,[36] and a specific focus on breaking with "the past," that is, traditional forms of obligation and religious practice.[37] And the backlash of this transformative lifestyle is socio-spiritual affliction.

If transnational Ghanaian Christians—even successful ones—fully divorced themselves from their extended families, then perhaps

---

35. Among enchanted Ghanaian Presbyterians, however, not all kin are destructive; extended Christian families have played a role in the development of enchanted Calvinism both in Ghana and in North America. For instance, the primary deliverance practitioner within the Presbyterian Church of Ghana, Catechist Ebenezer Abboah-Offei, is the cousin of Edward Okyere, who helped to found the Scripture Union Prayer Warriors. Scripture Union in turn helped to form and train the Presbyterian Church of Ghana's BSPG. And it was networks of kin—Abboah-Offei's wife is the niece of the wife of Samuel Atiemo, the pastor of the Brooklyn-based Ghanaian Presbyterian Church—that helped to establish the New York school of deliverance, which has become the primary institution for training deliverance practitioners in North America. Therefore, networks of Christian kin can also facilitate the spread of enchanted Christianity, providing solutions to conflicted kinship.

36. The congregation itself can function as a "new family," serving social security functions that have previously been provided by kinship groups in some circumstances and by welfare states in other circumstances. Joining a church enables converts at times to extricate themselves from old forms of social reciprocity and obligation (Marshall, *Political Spiritualities*, 194). In the creation of the BSPG in the 1960s, the Presbyterian leadership suggested that, because modern life breaks kinship ties and thrusts people into situations of great insecurity, the spiritual care of individuals needs to be addressed via this new organization and its healing practices (Presbyterian Church of Ghana, "Appendix F," 51).

37. Engelke, "Discontinuity"; Meyer, "Make a Complete Break."

these afflictions would decrease. But this is not the case. These Christians care tremendously for their kin in Ghana and their "hometown," or matrilineal lands, to which they belong. Most Ghanaians in the diaspora send remittances home—$1.6 billion in 2011—to various family members and their extended families, but particularly parents and siblings.[38] Most Ghanaian Presbyterians wish to build houses for themselves on communal lands—a sign of their success as migrants—and all (that I have ever spoken too) plan on being buried in Ghana, and typically on ancestral lands. Pastor Ofosu-Donkoh has, on more than one occasion, expressed his desire to become an ancestor: a departed family member that continues to play an active role in family affairs, even after his near death experience as a child at the hands of his maternal relatives. David Frimpong, after overcoming the spiritual attack instigated by his sister-in-law, did not retaliate but instead prayed for her forgiveness and continues to speak with her. Therefore, transnational migration, capitalism, and Christianity have not led most Ghaianians to fully break all ties with their extended families.[39] These social concepts of identity and belonging, to which obligation and reciprocity are attached, are partially embedded in the state apparatus via the customary land tenure system.

While biomedicine has had little effect on disenchanting Ghanaian Calvinism, the state in Ghana has contributed to its enchantment in the form of low welfare spending as well as the maintenance of the customary land tenure system. The state in Ghana, however, has made significant improvements over the last ten years with welfare spending, including instituting a nationwide health insurance scheme, but it is certainly not able to offer the social safety net of European states or even the United States. And in the United States, Ghanaian Presbyterians, some of whom are illegal, do not have great access to welfare opportunities, which are significantly less than in Europe. The state in Ghana also affects the quality that enchantment takes among Ghanaian Presbyterians; the socio-spiritual afflictions that are the backlash of avoiding kinship-based expectations of redistribution are embedded in the state apparatus within the customary land tenure system. These socio-spiritual afflictions affect Ghanaian

---

38. Dogbevi, "Remittances"; Mazzucato, van den Boom, and Nsowah-Nuamah, "Origin and Destination," 144.

39. See Engelke for his treatment of the "demands of the break" in the everyday lives of the Friday Apostolics of Zimbabwe ("Past Pentecostalism," 187–95).

Presbyterians in both Ghana and the United States. Such conditions often engender enchanted Christianity.

Max Weber proclaimed the disenchantment of Christianity more than one hundred years ago. More recently, Jean-Luc Nancy describes Christianity as so disenchanted that, in fact, it is a form of atheism.[40] Nancy writes that the plurality of gods within polytheism corresponds to their effective presence in the lives of believers in terms of power, threat, and assistance. Alternatively, the monotheistic God signifies the withdrawal of God's presence as well as those corresponding power relationships. In Nancy's juxtaposition between polytheism and monotheism, one can see the similar employment of ideal types that Kant utilizes when making his distinction between religion as cult, which seeks material benefits such as health from God, versus religion as moral action (i.e., Protestantism) that commands human beings to lead better lives through behavior changes. But as I demonstrate in the introduction of this book, ideal types are problematic and neglectful of historic and ethnographic realities.

Today, more than a quarter of the world's 2.2 billion Christians identify themselves as Pentecostals or Charismatics, and the practice of divine healing distinguishes them, more than any other practice, from other Christians.[41] This popular form of enchanted Christianity—which includes the transnational Ghanaian Presbyterian community that is the subject of this book—focuses distinctly on *both* healing and success in daily life, as well as on industriousness and changing day-to-day behaviors to lead better lives. In essence, forms of popular Christianity throughout the world are hybrids of these opposing ideal types described by Kant, Weber, and Nancy. These creative forms of hybridization within Pentecostal and Charismatic Christianity allow this faith to serve the needs of the suffering that fill its churches. And in the end, Jacques Derrida reminds us that ideal types do not exist; all discourse and practice—including Christian discourse and practice—is syncretism or *bricolage*.[42]

---

40. Nancy, *Dis-Enclosure*, 35.
41. Lugo et al., *Spirits and Power*.
42. Derrida, *Writing and Difference*, 285.

Appendix

Deliverance Questionnaire

Grace Evangelistic Team
P. O. Box 299
Akropong—Akuapim

File No. _____   Date _____

GRACE EVANGELISTIC TEAM
QUESTIONNAIRE FOR DETECTING PERSONS WHO NEED SPIRITUAL MINISTRATION
ANSWER "YES" OR "NO" GIVE HONEST DETAILS

PERSONAL PARTICULARS   STRICTLY CONFIDENTIAL

Name._____
Meaning of name _____
Tribe _____ Clan _____ Hometown _____
Usual place of residence _____ Tel. _____
Occupation _____ Sex: [ ] Male [ ] Female Age _____
Marital status: [ ] Single [ ] Married [ ] Widowed [ ] Co-habiting. No. of Children Alive_____ Dead _____
Position in family tree (mother line) _____ out of _____
Are you born again? _____ When (year)? _____ Where? _____

PRESENT COMPLAINTS AND DURATION (I.E. PRESENT PROBLEM)
(State your problem(s) below)
_____
_____
_____
_____
_____

How long have you been in this problem? _____
What have you done about the problem(s)? _____
Have you gone through deliverance ministry before? _____
When _____ Where _____
What happened? _____

FAMILY BACKGROUND (SPIRITUAL EXPOSURES)
Parents' church or religion: a. Father's _____ b. Mother's _____
Have they any connection with a church where all the members wear the same dress (e.g., Aladura). [ ] Yes. [ ] No.
Name them _____
Have your parents made any visit(s) to native doctor(s)? [ ] Yes. [ ] No. [ ] I don't know.
Were there any incisions (i.e., cuts) made to rub black powder (mmoto) into their bodies?
    [ ] Yes. [ ] No. [ ] I don't know.
Did they offer any sacrifice? [ ] Yes. [ ] No. [ ] I don't know.

If yes, state the nature of the sacrifice(s). _____
Do your parents belong to any transcendental meditation groups, e.g., lodges. [ ] Yes. [ ] No. [ ] I don't know.
Which group(s)? _____
Have your parents any ancestral shrine(s) or god(s)? [ ] Yes. [ ] No. [ ] I don't know.
If yes, what is the name of the god or shrine? _____
What does the shrine/god dislike (taboos?) _____
Have your parents any ancestral stools? Fathers' line: [ ] Yes. [ ] No. Mothers' line: [ ] Yes. [ ] No.
Do you know of any covenant or curse on you or your family? [ ] Yes. [ ] No. [ ] I don't know.
If yes, state it (them). _____

PERSONAL DETAILS

Are you twin born? [ ] Yes. [ ] No.
Your church or religion _____
Do you have any connection with a church where they wear the same dress? [ ] Yes. [ ] No.
Do you have connection with some other prayer houses? [ ] Yes. [ ] No.
If yes, name them _____
Have you made any visits to native doctors? [ ] Yes. [ ] No.
Have any incisions (cuts) been made on any part of your body? [ ] Yes. [ ] No.
Have you ever offered any sacrifice(s) to any god(s)? [ ] Yes. [ ] No.
Have you any contacts with dwarfs? [ ] Yes. [ ] No.
Have you ever made any visit(s) to a mallam? [ ] Yes. [ ] No.
If yes, what happened? _____
Have you ever taken any black powder (mmoto)? [ ] Yes. [ ] No.
Have you ever drunk any concoction(s)? [ ] Yes. [ ] No.
Have you taken a concoction bath? [ ] Yes. [ ] No.
Have you used any concoction or talisman for (a) trading? [ ] Yes. [ ] No. (b) travelling? [ ] Yes. [ ] No.
(c) Boys/girls? [ ] Yes. [ ] No. (d) Stealing? [ ] Yes. [ ] No. (e) protection? [ ] Yes. [ ] No.
Have you used any concoction/talisman/charms for any other purpose other than those stated above?[ ] Yes. [ ] No.
If yes, give details? _____
Do you know any contributory story told you by your parents or relatives in relation to your birth and childhood?
    [ ] Yes. [ ] No.
    Give details, if any _____
Who did you stay with as a child? _____
Do you have any history of rejection in childhood? [ ] Yes. [ ] No.
Give details, if any _____
Were you (a) convulsive: [ ] Yes [ ] No; (b) sickly: [ ] Yes [ ] No during childhood?

PERSONAL STRANGE PHENOMENA

Do you "see" (hallucinations) things, which are not really there?	[ ] Yes.	[ ] No.
Do you "hear" (auditory voices) strange voices when nothing is around?	[ ] Yes.	[ ] No.
Do you "smell" (olfactory) things, which are not present?	[ ] Yes.	[ ] No.
Do you know things before they happen?	[ ] Yes.	[ ] No.
Do you have additional money or other items you do not know they came about?	[ ] Yes.	[ ] No.
Do you lose your money or other items often?	[ ] Yes.	[ ] No.
Have you worn any rings/clothes/bangles you never knew how they came?	[ ] Yes.	[ ] No.
Have you lost your wedding/engagement ring mysteriously?	[ ] Yes.	[ ] No.
Do you chat with strange voices from within you?	[ ] Yes.	[ ] No.
Do you have a feeling of invisible presence when you are alone?	[ ] Yes.	[ ] No.
How do you feel when you hear traditional drum music?	[ ] Yes.	[ ] No.

PERSONAL STRANGE CHARACTERISTICS

Do you have excessive (too much)(a) anger? [ ] Yes [ ] No; (b) hatred? [ ] Yes [ ] No;
(c) Fear of snakes? [ ] Yes [ ] No. (d) fear of water (sea, river, ponds, etc.)?[ ] Yes. [ ] No.
(e) Are you easily scared? [ ] Yes. [ ] No.
Do you have sex with anybody who wants to (more than one sexual partner/indiscriminate sex)? [ ] Yes. [ ] No
Do you do any of the following sexual perversions, i.e., abnormal sexual behaviours?
a) Masturbation [ ] Yes. [ ] No.
b) Homosexuality/Lesbianism (i.e., sexually attracted to people of your own sex - man to man or woman to woman
[ ] Yes. [ ] No. c) Sex with animals (bestiality). [ ] Yes. [ ] No.
Do you indulge in any other sexual perversions apart from those referred above? [ ] Yes. [ ] No.
Do you have suicidal thoughts (i.e., do you sometimes think of killing yourself?)? [ ] Yes. [ ] No.
Do you like weeping always? [ ] Yes. [ ] No.
Are you not serious over situations considered serious by others and laugh unnecessarily? [ ] Yes. [ ] No
Do you want something and at the same time you feel you don't need it? [ ] Yes. [ ] No
Are you very stubborn? [ ] Yes. [ ] No.
Do you enjoy seeing others suffer? [ ] Yes. [ ] No.
Are you restless? [ ] Yes. [ ] No.
Do you like these things so much that you cannot do without them? (a) Alcohol [ ] Yes. [ ] No.
b) Drugs [ ] Yes. [ ] No. (c) Sleep [ ] Yes. [ ] No. (d) Food [ ] Yes. [ ] No.
(d) Chewing gum [ ] Yes. [ ] No. (e) Outrageous/Excessive dressing [ ] Yes. [ ] No.
Any other? _____
Are you quarrelsome? [ ] Yes. [ ] No. Do you procrastinate a lot? [ ] Yes. [ ] No.
Do you feel unnatural heat movement in your body? [ ] Yes. [ ] No.
Do you experience terrible menstrual period (females only) [ ] Yes. [ ] No.
Do you experience any of the following? (a) Sleepless night [ ] Yes. [ ] No.
(b) biting of fingernails [ ] Yes. [ ] No. (e) Scratching or fidgeting with your body unnaturally [ ] Yes. [ ] No
Do you experience laziness?[ ] Yes. [ ] No. Do you experience dizziness? [ ] Yes. [ ] No.
Do you experience failure in your life endeavours? [ ] Yes. [ ] No.
Do you get sad and moody/depressed without cause? [ ] Yes. [ ] No.
Are you easily irritated (become angry or annoyed)? [ ] Yes. [ ] No.
Do you get forget easily, i.e., memory loss? [ ] Yes. [ ] No.

Are you afraid of mixing with people, hence desiring to be alone? [ ] Yes. [ ] No.
Do you have a negative self-image, i.e., do you think of yourself as a failure or never do well? [ ] Yes. [ ] No.

PECULIAR DREAM STATE
Concerning your dreams, do the following happens to you in the dreams?
Do you forget your dreams? [ ] Yes. [ ] No.
Do you see yourself picking mushrooms or snails? [ ] Yes. [ ] No.
Do you have nightmares, e.g., that you are being chased by masquerades (people or things wearing masks so that you cannot see their faces) or animals, or do you find yourself falling off cliffs or mountains into deep ditches or pits, etc.??
[ ] Yes. [ ] No.
Do you find yourself always in your childhood house, wearing old clothes (school uniform) or at former work place or school? [ ] Yes. [ ] No.
Do you see yourself being burgled, trying to locate lost property, your goods being auctioned?? [ ] Yes. [ ] No.
Do you see yourself handling huge sums of money? [ ] Yes. [ ] No.
Do you see yourself locked up in a room unable to come out or handcuffed, chained or imprisoned? [ ] Yes. [ ] No.
Do you have sexual affair with somebody in your dream? [ ] Yes. [ ] No.
Do have a feeling of being pressed down and unable to talk? [ ] Yes. [ ] No.
Do you attend any regular parties or meetings? [ ] Yes. [ ] No.
Do you see yourself doing menial jobs far below your status in real life? [ ] Yes. [ ] No.
Do you fight? [ ] Yes. [ ] No.
Do you take part in a wedding/marriage ceremony of yourself? [ ] Yes. [ ] No.
Do you see the following before your menstrual period? (Females only) red oil [ ] Yes. [ ] No.
ripe fruits [ ] Yes. [ ] No. fresh meat with blood [ ] Yes. [ ] No. sexual intercourse [ ] Yes. [ ] No.
Do you see yourself pregnant? [ ] Yes. [ ] No. Giving birth? [ ] Yes. [ ] No.
Carrying babies? [ ] Yes. [ ] No. before your period before it actually comes?
Do you swim in water (sea, lake, river, etc.)? [ ] Yes. [ ] No.
Do you play with age mates in water? [ ] Yes. [ ] No; stand by a bank or shore or a river, sea? [ ] Yes. [ ] No; always crossing with a canoe [ ] Yes. [ ] No;
crossing a bridge which breaks midway, falling headlong in into the river? [ ] Yes. [ ] No.
Do you find yourself by large rivers you cannot cross? [ ] Yes. [ ] No;
climbing staircases and missing your steps or ladders with missing steps? [ ] Yes. [ ] No.
Do you find yourself by a big gutter or trench you cannot cross, or in a situation where you cannot go forward?
[ ] Yes. [ ] No.
Do you see or play with snakes? [ ] Yes. [ ] No.
Do you find yourself wandering in the forest, i.e., walking around without any clear purpose or direction?
[ ] Yes. [ ] No.
Do you see yourself flying (levitation)? [ ] Yes. [ ] No.
Do you go to a specific market? [ ] Yes. [ ] No.
Do you see yourself at a rubbish dump? [ ] Yes. [ ] No.
Do you find yourself eating? [ ] Yes. [ ] No.
Do you see, talk with dead friends/relatives (necromancy) [ ] Yes. [ ] No;
Do you climbing a mountain or hill and never getting to the top? [ ] Yes. [ ] No.
Do you see strange animals like cats, ants, mice, dogs, etc.? [ ] Yes. [ ] No.

Is there any information that the questionnaire did not ask, but which you feel might be of some help that you wish to give? Please state them. _____
_____
_____ (You may continue on another sheet.)

INVESTIGATION AND RESULTS (FOR OFFICE USE ONLY)

Name of Investigator _____

Remarks _____
_____
_____
_____

Bibliography

Abboah-Offei, Ebenezer. *Church Leaders Training Manual: Practical Theology Resource in the School of Deliverance.* Akropong: Grace Presbyterian Church, 2006.

Ackah, Charles, and Jann Lay. "Gender Impacts of Agricultural Liberalization: Evidence from Ghana." In *Gender Aspects of the Trade and Poverty Nexus: A Macro-Micro Approach,* edited by Maurizio Bussolo and Rafael E. De Hoyos, 217–45. New York: Palgrave Macmillan, 2009.

Adogame, Afe, and Cordula Weisskoppel. "Introduction: Locating Religion in the Context of African Migration." In *Religion in the Context of African Migration,* edited by Afe Adogame and Cordula Weisskoppel, 1–22. Bayreuth: Pia Thielmann and Eckhard Breitinger, 2005.

Agha, Asif. "The Speciation of Modernity." Discussant comments on papers presented in the panel "Linguistic Modernity and Its Discontents: Mixed Evidence, Hybrid Models," 99th Annual Meeting, American Anthropological Association, San Francisco, November 17, 2000.

Aiken, Linda H. "U.S. Nurse Labor Market Dynamics Are Key to Global Nurse Sufficiency." *Health Service Research* 42, no. 3, part 2 (2007): 1299–1320.

Akyeampong, Emmanuel, and Pashington Obeng. "Spirituality, Gender, and Power in Asante History." *International Journal of African Historical Studies* 28, no. 3 (1995): 481–508.

Allman, Jean, and John Parker. *Tongnaab: The History of a West African God.* Bloomington: Indiana University Press, 2005.

Allman, Jean, and Victoria Tashjian. *"I Will Not Eat Stone": A Women's History of Colonial Asante.* Portsmouth, NH: Heinemann, 2000.

Anim, Peter. *The History of How the Full Gospel Church Was Founded in Ghana.* Accra: Graphic Press, n.d.

Ankins, John Wesley. *Words of Healing.* Philadelphia: Faith Tabernacle Publishing House, 1908.

Appiah-Kubi, Kofi. *Man Cures, God Heals: Religion and Medical Practice among the Akans of Ghana.* New York: Friendship Press, 1981.

Arhin, Kwame. "The Political and Military Roles of Akan Women." In *Female and Male in West Africa,* edited by Christine Oppong, 91–96. London: Allen and Unwin, 1983.

Arku, Cynthia, and Frank S. Arku. "I Cannot Drink Water on an Empty Stomach: A Gender Perspective on Living with Drought." *Gender and Development* 18, no. 1 (2010): 115–24.

———. "More Money, New Household Cultural Dynamics: Women in Micro-Finance in Ghana." *Development in Practice* 19, no. 2 (2009): 200–213.
Asamoah-Gyadu, J. Kwabena. *African Charismatics: Current Developments within Independent Indigenous Pentecostalism in Ghana.* Leiden: Brill, 2005.
Austrian Red Cross. *Health Care in Ghana.* Vienna: ACCORD, 2009.
Awal, Mohammed. "Ghana: Democracy, Economic Reform, and Development, 1993–2008." *Journal of Sustainable Development in Africa* 14, no. 1 (2012): 97–118.
Babou, Cheikh Anta. "Migration and Cultural Change: Money, 'Caste,' Gender, and Social Status among Senegalese Female Hair braiders in the United States." *Africa Today* 55, no. 2 (2008): 3–22.
Baeta, Christian G. *Prophetism in Ghana: A Study of Some "Spiritual" Churches.* London: SCM Press, 1962.
Barker, Peter, and Samuel Boadi-Siaw. *Changed by the Word: the Story of Scripture Union Ghana.* Bangalore: Bangalore Offset Printers, 2005.
Barry, John M. *The Great Influenza: The Epic Story of the Deadliest Plague in History.* New York: Viking, 2004.
Bartels, Francis L. *The Roots of Ghana Methodism.* Cambridge: Cambridge University Press, 1965.
Beeko, Anthony A. *The Trail Blazers: Fruits of 175 Years of the Presbyterian Church of Ghana (1828–2003).* Accra: Afram Publications, 2004.
Berry, Sarah. *Cocoa, Custom, and Socio-Economic Change in Rural Western Nigeria.* Oxford: Clarendon Press, 1975.
———. *No Condition Is Permanent: The Social Dynamics of Agrarian Change in Sub-Saharan Africa.* Madison: University of Wisconsin Press, 1993.
Biney, Moses. *From Africa to America: Religion and Adaptation Among Ghanaian Immigrants in New York.* New York: New York University Press, 2011.
———. "'Singing the Lord's Song in a Foreign Land': Spirituality, Communality, and Identity in a Ghanaian Immigrant Congregation." In Olupona and Gemignani, *African Immigrant Religions in America*, 259–78.
Boardman, William E., ed. *Record of the International Conference on Divine Healing and True Holiness.* London: J. Snow, 1885.
Boddy, Janice. *Wombs and Alien Spirits: Women, Men, and The Zar Cult in Northern Sudan.* Madison: the University of Wisconsin Press, 1989.
———. "Managing Tradition: Superstition and the Making of National Identity among Sudanese Women Refugees." In *The Pursuit of Certainty: Religious and Cultural Formulations*, edited by Wendy James, 15–44. London: Routledge, 1995.
Bolzendahl, C. I. and D. J. Myers. "Feminist attitudes and support for gender equality: opinion change in women and men, 1974–1998." *Social Forces* 83, no. 2 (2004): 759–89.

Bourdieu, Pierre. "The Social Space and the Genesis Groups." *Theory and Society* 14, no. 6 (1985): 723–44.
Bourguingnon, Erika. "Suffering and Healing, Subordination and Power: Women and Possession Trance." *Ethnos* 32, no. 4 (2004): 557–74.
Bowdich, T. Edward. *Mission from Cape Coast Castle to Ashantee*. 1819. London: Frank Cass, 1966.
Brahinsky, Josh. "Pentecostal Body Logics: Cultivating a Modern Sensorium." *Cultural Anthropology* 27, no. 2 (2012): 215–38.
Bredwa-Mensah, Yaw. "The Church of Pentecost in Retrospect: 1937–1960." In *James McKeown Memorial Lectures*, 1–27. Accra: The Church of Pentecost, 2004.
Brokensha, David. *Social Change at Larteh, Ghana*. Oxford: Clarendon Press, 1966.
Brown, Candy Gunther. "Introduction: Pentecostalism and the Globalization of Illness and Healing." In Brown, *Global Pentecostal and Charismatic Healing*, 3–26.
Brown, Candy Gunther, ed. *Global Pentecostal and Charismatic Healing*. New York: Oxford University Press, 2011.
Brown, Karen McCarthy. *Mama Lola: A Vodou Priestess in Brooklyn*. Berkeley and Los Angeles: University of California Press, 2001.
Brydon, Lynne. "Ghanaian Responses to the Nigerian Expulsions of 1983." *African Affairs* 84, no. 337 (1985): 561–85.
Bussolo, Maurizio, and Denis Medvedev. *Challenges to MDG Achievement in Low Income Countries: Lessons from Ghana and Honduras*. Washington, DC: The World Bank Development Economics Prospects Group, 2007.
Cannell, Fenella. "Introduction." In *The Anthropology of Christianity*, edited by Fenella Cannell, 1–50. Durham: Duke University Press, 2007.
Capone, Stefania. *Searching for Africa in Brazil: Power and Tradition in Candomblé*. Durham: Duke University Press, 2010.
Caps, Randy, Kristen McCabe, and Michael Fix. *New Streams: Black African Migration to the United States*. Washington, DC: Migration Policy Institute, 2011.
Chestnut, R. Andrew. *Born Again in Brazil: The Pentecostal Boom and the Pathogens of Poverty*. New Brunswick: Rutgers University Press, 1997.
Christaller, Johann G. *Dictionary of the Asante and Fante Language Called Tshi (Twi)*. 1881. Basel: Basel Evangelical Missionary Society, 1933.
Clark, Mary Ann. *Where Men Are Wives and Mothers Rule: Santeria Ritual Practices and Their Gender Implications*. Gainesville: University of Florida Press, 2004.
Cohen, Emma. "What Is Spirit Possession?: Defining, Comparing, and Explaining Two Possession Forms." *Ethnos* 73, no. 1 (2008): 101–26.
Cole, Catherine M., Takyiwaa Manuh, and Stephan F. Miescher, eds. *Africa After Gender?* Bloomington: Indiana University Press, 2007.

Colleyn, Jean-Paul. "Horse, Hunter, and Messenger: The Possessed Men of the Nya Cult in Mali." In *Spirit Possession: Modernity and Power in Africa*, edited by Heike Behrend and Ute Luig, 68–78. Madison: University of Wisconsin Press, 1999.

Comaroff, John L., and Jean Comaroff. *Of Revelation and Revolution.* Vol. 1, *Christianity, Colonialism, and Consciousness in South Africa.* Chicago: University of Chicago Press, 1991.

———. *Of Revelation and Revolution.* Vol. 2, *The Dialectics of Modernity on a South African Frontier.* Chicago: University of Chicago Press, 1997.

Cox, Harvey. *Fire from Heaven: The Rise of Pentecostal Spirituality and the Reshaping of Religion in the Twenty-first Century.* Cambridge, MA: Da Capo Press, 1995.

Crabtree, Steve, and Brett Pelham. "What Alabamians and Iranians Have in Common: A Global Perspective on Americans' Religiosity Offers a Few Surprises." *Gallup.com*, February 9, 2009. www.gallup.com/poll/114211/alabamians-iranians-common.aspx.

Curtin, Philip D. "Medical Knowledge and Urban Planning in Colonial Tropical Africa." In Feierman and Janzen, *Social Basis of Health and Healing in Africa*, 235–55.

———. *Migration and Mortality in Africa in the Atlantic World, 1700–1900.* Burlington, VT: Ashgate Variorum, 2001.

Curtis, Heather. *Faith in the Great Physician: Suffering and Divine Healing in American Culture, 1860–1900.* Baltimore: Johns Hopkins University Press, 2007.

Danker, William J. *Profit for the Lord: Economic Activities in Moravian Missions and the Basel Mission Trading Company.* Grand Rapids, MI: Eerdmans, 1971.

Daswani, Girish. "Transformation and Migration among Members of a Pentecostal Church in Ghana and London." *Journal of Religion in Africa* 40, no. 4 (2010): 424–27.

Debrunner, Hans. *A History of Christianity in Ghana.* Accra: Waterville, 1967.

———. *Witchcraft in Ghana: A Study on the Belief in Destructive Witches and Its Effect on the Akan Tribes.* Accra: Presbyterian Book Depot, 1961.

Deininger, Klaus. *Land Policies for Growth and Poverty Reduction.* Washington, DC: The World Bank, 2003.

Derrida, Jacques. *Writing and Difference.* Chicago: University of Chicago Press, 1978.

Dilger, Hansjorg, Abdoulaye Kane, and Stacey A. Langwick, eds. *Medicine, Mobility, and Power in Global Africa: Transnational Health and Healing.* Bloomington: Indiana University Press, 2013.

Dittman, Emily. "A Pentecostal African Church as a Support Network: The Church of Pentecost, Philadelphia Assembly." BA thesis, University of Pennsylvania, 2003.

Dogbevi, Emmanuel K. "Remittances from Ghanaians Abroad Outstrip Overseas Dev't Assistance," Ghanaweb. September 13, 2011.

Dovlo, Delanyo. "Migration of Nurses from Sub-Saharan Africa: A Review of Issues and Challenges." *Health Service Research* 42, no. 3, part 2 (2007): 1373–88.

Ellis, Alfred B. *The Tshi-Speaking Peoples of the Gold Coast of West Africa.* 1887. Chicago: Benin Press, 1964.

Ellis, Stephen. *The Mask of Anarchy: The Destruction of Liberia and the Religious Dimension of an African Civil War.* 2nd ed. New York: New York University Press, 2007.

Engelke, Matthew. "Discontinuity and the Discourse of Conversion." *Journal of Religion in Africa* 34, nos. 1–2 (2004): 82–109.

———. "Past Pentecostalism: Notes on Rupture, Realignment, and Everyday Life in Pentecostal and African Independent Churches." *Africa* 80, no. 2 (2010): 177–99.

ESODEC-UNICEF. *Analysis of the 2010 National Budget and Economic Policy Statement of the Government of Ghana to Determine Gaps and Opportunities for Women and Children.* March 2010.

Etherington, Norman. "Missionary Doctors and African Healers in Mid-Victorian South Africa." *South African Historical Journal* 19 (1987): 77–91.

Evans-Pritchard, E. E. "Sorcery and Native Opinion." *Africa* 4, no. 1 (1931): 23–28.

Fannon, Franz. *A Dying Colonialism.* New York: Grove Press, 1967.

Feierman, Steven. "Explanation and Uncertainty in the Medical World of Ghaambo." *Bulletin of the History of Medicine* 74, no. 2 (2000): 317–44.

———. "Healing as Social Criticism in the Time of Colonial Conquest." *African Studies* 54, no. 1 (1995): 73–88.

———. "Therapy as a System-in-Action in Northwestern Tanzania." *Social Science and Medicine* 15B, no. 3 (1981): 353–60.

Feierman, Steven, and John M. Janzen. "Introduction." In Feierman and Janzen, *Social Basis of Health and Healing in Africa*, 1–24.

Feierman, Steven, and John J. Janzen, eds. *The Social Basis of Health and Healing in Africa.* Berkeley and Los Angeles: University of California Press, 1992.

Field, Margaret J. *Search for Security: An Ethno-Psychiatric Study of Rural Ghana.* New York: W. W. Norton, 1960.

Fischer, Friedrich H. *Der Missionsarzt Rudolf Fisch und die Anfange Medizinischer Arbeit der Basler Mission an der Goldkuste (Ghana).* Herzogenrath: Murken-Altrogge, 1991.

Flory, Margaret. *Dear House, Mission Becomes You: Gilmor Sloane House Stony Point Center Stony Point, New York 1949–1999.* Louisville: Bridge Resources, 2000.

Foucault, Michel. *Abnormal.* New York: Picador, 2003.
———. *Birth of the Clinic: An Archaeology of Medical Perception.* New York: Vintage Books, 1994.
Frimpong-Ansah, Jonathan H. *The Vampire State in Africa: The Political Economy of Decline in Ghana.* London: James Currey, 1992.
Fumanti, Mattia. "'Virtuous Citizenship': Ethnicity and Encapsulation among Akan-Speaking Ghanaian Methodists in London." *African Diaspora* 3, no. 1 (2010): 13–42.
Fyfe, Christopher. *A History of Sierra Leone.* London: Oxford University Press, 1962.
Gaudio, Rudolf Pell. *Allah Made Us: Sexual Outlaws in an Islamic African City.* Malden, MA: Wiley-Blackwell, 2009.
Gelfand, Michael. *Livingstone the Doctor, His Life and Travels: A Study in Medical History.* Oxford: Blackwell, 1957.
Gemignani, Regina. "Gender, Identity, and Power in African Immigrant Evangelical Churches." In Olupona and Gemignani, *African Immigrant Religions in America,* 133–57.
Geschiere, Peter. *The Modernity of Witchcraft: Politics and the Occult in Postcolonial Africa.* Charlottesville: University of Virginia Press, 1997.
Gifford, Paul. *African Christianity: Its Public Role.* Bloomington: Indiana University Press, 1998.
———. *Ghana's New Christianity: Pentecostalism in a Globalizing African Economy.* Bloomington: Indiana University Press, 2004.
Gill, Anthony, and Erik Lundsgaarde. "State Welfare Spending and Religiosity: A Cross-National Analysis." *Rationality and Society* 16, no. 4 (2004): 399–436.
Gloege, Timothy. "The Moody Bible Institute and Pentecostalism (1889–1930): Fundamentalist-Pentecostal Conflicts and BMI's Crisis of Function." MA thesis, Wheaton College, 2000.
Gray, Natasha. "Independent Spirits: The Politics of Policing Anti-Witchcraft Movements in Colonial Ghana, 1908–1927." *Journal of Religion in Africa* 35, no. 2 (2005): 139–58.
———. "Witches, Oracles, and Colonial Law: Evolving Anti-Witchcraft Practices in Ghana, 1927–1932." *International Journal of African Historical Studies* 34, no. 2 (2001): 339–63.
Green, Jeffrey E. "Two Meanings of Disenchantment: Sociological Condition vs. Philosophical Act—Reassessing Max Weber's Thesis of the Disenchantment of the World." *Philosophy and Theology* 17, nos. 1–2 (2005): 51–84.
Green, Maia. "Medicine and the Embodiment of Substances among Pogoro Catholics, Southern-Tanzania." *Journal of the Royal Anthropological Institute* 2, no. 3 (1996): 1–14.

Green, Sandra E. *Sacred Sites and Colonial Encounter: A History of Meaning and Memory in Ghana.* Bloomington: Indiana University Press, 2002.

Grier, Beverly C. "Pawns, Porters, and Petty Traders: Women in the Transition to Cash Crop Agriculture in Colonial Ghana." *Signs* 17, no. 2 (1992): 304–29.

Grundmann, Christoffer H. *Sent to Heal!: Emergence and Development of Medical Missions.* Lanham, MD: University Press of America, 2005.

Guest, William. *Pastor Blumhardt and His Work.* London: Morgan and Scott, 1881.

Harakas, Stanley Samuel. *Health and Medicine in the Eastern Orthodox Tradition.* New York: Crossroad, 1990.

Hardesty, Nancy. *Faith Cure: Divine Healing in the Holiness and Pentecostal Movements.* Peabody, MA: Hendrickson, 2003.

Hardiman, David, ed. *Healing Bodies, Saving Souls: Medical Missions in Asia and Africa.* Amsterdam: Rodopi, 2006.

Harris, Hermione. *Yoruba in Diaspora: An African Church in London.* New York: Palgrave Macmillan, 2006.

Hecker, Daniel E. "Occupational Employment Projections to 2014." *Monthly Labor Review* 128, no. 11 (2005): 70–101.

Hennock, E. P. *The Origins of the Welfare State in England and Germany, 1850–1914: Social Policies Compared.* New York: Cambridge University Press, 2007.

Hill, Polly. *The Gold Coast Cocoa Farmer: A Preliminary Survey.* London: Oxford University Press, 1956.

———. "The Migrant Cocoa Farmers of Southern Ghana." *Africa* 31, no. 3 (1961): 209–30.

———. *The Migrant Cocoa-Farmers of Southern Ghana: A Study in Rural Capitalism.* Cambridge: Cambridge University Press, 1963.

Hochschild, Arlie. "Love and Gold." In *Global Woman: Nannies, Maids, and Sex Workers in the New Economy,* edited by B. Ehrenreich and A. Hochschild, 15–55. New York: Metropolitan, 2003.

Hodzic, Saida. "Unsettling Power: Domestic Violence, Gender Politics, and Struggles over Sovereignty in Ghana." *Ethnos* 74, no. 3 (2009): 331–60.

Hokkanen, Markku. *Medicine and Scottish Missionaries in the Northern Malawi Region, 1875–1930: Quest for Health in a Colonial Society.* Lewiston, NY: The Edwin Mellen Press, 2007.

Hyden, Goran. *Byond Ujamaa in Tanzania: Underdevelopment and an Uncaptured Peasantry.* Berkeley and Los Angeles: University of California Press, 1980.

Igreja, Victor, Beatrice Dias-Lambranca, and Annemiek Richters. "*Gamba* Spirits, Gender Relations, and Healing in Post–Civil War Gorongosa, Mozambique." *Journal of the Royal Anthropological Institute* 14, no. 2 (2008): 353–71.

Ising, Dieter. *Johann Christoph Blumhardt, Life and Work.* Eugene, OR: Cascade Books, 2009.

Janzen, John. "Afri-Global Medicine: New Perspectives on Epidemics, Drugs, Wars, Migrations, and Healing Rituals." In Dilger, Kane, and Langwick, *Medicine, Mobility, and Power in Global Africa*, 115–37.

Jenkins, Paul. "Afterward: The Basel Mission, the Presbyterian Church, and Ghana since 1918." In *Missionary Zeal and Institutional Control: Organizational Contradictions in the Basel Mission on the Gold Coast, 1828–1917*, by Jon Miller, 195–222. Grand Rapids, MI: William B. Eerdmans, 2003.

———. "The Church Missionary Society and the Basel Mission: An Early Experiment in Inter-European Cooperation." *The Church Mission Society and World Christianity, 1799–1999*, edited by Kevin Ward and Brian Stanley, 43–65. Grand Rapids, MI: William B. Eerdmans, 2000.

———. "The Scandal of Continuing Intercultural Blindness in Mission Historiography: The Case of Andreas Riis in Akwapim." *International Review of Missions* 87, no. 344 (2007): 67–76.

———."Villagers as Missionaries: Württemberg Pietism as a Nineteenth-Century Missionary Movement." *Missiology: An International Review* 8, no. 4 (1980): 425–32.

Jenkins, Philip. *The Next Christendom: The Coming of Global Christianity.* Oxford: Oxford University Press, 2002.

Jennings, Michael. "'Healing of Bodies, Salvation of Souls': Missionary Medicine in Colonial Tanganyika, 1870s–1939." *Journal of Religion in Africa* 38, no. 1 (2008): 27–56.

Kamenetsky, Christa. *The Brothers Grimm and Their Critics: Folktales and the Quest for Meaning.* Athens: Ohio University Press, 1992.

Kant, Immanuel. *Religion within the Boundaries of Mere Reason.* 1763. Cambridge: Cambridge University Press, 1998.

Karkkainen, Veli-Matti."The Pentecostal View." In *The Lord's Supper: Five Views*, edited by Gordon T. Smith, 117–44. Downers Grove, IL: IVP Academic, 2008.

Kay, Geoffrey B. *The Political Economy of Colonialism in Ghana.* Cambridge: Cambridge University Press, 1972.

Keane, Webb. "Religious Language." *Annual Review of Anthropology* 26 (1995): 47–71.

Kieran, John. "Some Roman Catholic Missionary Attitudes to Africans in Nineteenth-Century East Africa." *Race* 10, no. 3 (1969): 341–59.

Killick, Tony. *Development Economics in Action: A Study of Economic Policies in Ghana.* 2nd ed. New York: Routledge, 2010.

Kim, Kwang Chug, and Shin Kim. "Ethnic Role of Korean Immigrant Churches in the United States." In *Korean Americans and Their Religions: Pilgrims and Missionaries from a Different Shore*, edited by Ho-Youn Kwon,

Kwang Chung Kim, and R. Stephen Warner, 71–94. University Park: The Pennsylvania State University Press, 2001.

Kingma, Mireilee. "Nurses on the Move: A Global Overview." *Health Service Research* 42, no. 3, part 2 (2007): 1281–98.

Kofman, Eleonore. "Gendered Global Migrations: Diversity and Stratification." *International Feminist Journal of Politics* 6, no. 4 (2004): 643–65.

Kotin, Timothy. "Health Insurance in Ghana." *Harvard Political Review*, 2010. http://hpronline.org/health-insurance-in-ghana/.

Kumah, Paulina. *Twenty Years of Spiritual Warfare. Accra: Scripture Union Prayer Warriors Ministry.* Accra: SonLife Printing Press, 1994.

Kumi, Rev. T. A. "Appendix I: Supplement to Report on Ramseyer Memorial Retreat and Study Centre, Abetifi." In *Minutes of the 34th Synod*, 51–53. Akropong: Presbyterian Church of Ghana, 1963.

Langwick, Stacey. "Articulate(d) Bodies: Traditional Medicine in a Tanzanian Hospital." *American Ethnologist* 35, no. 3 (2008): 428–39.

Langwick, Stacey, Hansjorg Dilger, and Abdoulaye Kane. "Introduction: Transnational Medicine, Mobile Experts." In Dilger, Kane, and Langwick, *Medicine, Mobility, and Power in Global Africa*, 1–30.

Larbi, Emmanuel Kingsley. *Pentecostalism: The Eddies of Ghanaian Christianity.* Accra: Centre for Pentecostal and Charismatic Studies, 2001.

Lewis, Ioan M. *Ecstatic Religion: An Anthropological Study of Spirit Possession and Shamanism.* Harmondsworth, UK: Penguin, 1971.

Lipset, Seymour Martin. *The First New Nation: The United States in Comparative and Historical Perspective.* New York: W. W. Norton, 1963.

Ludwig, Frieder, and J. Kwabena Asamoah-Gyadu, eds. *African Christian Presence in the West: New Immigrant Congregations and Transnational Networks in North America and Europe.* Trenton, NJ: Africa World Press, 2011.

Lugo, Luis, Sandra Stencel, John Green, Timothy S. Shah, Brian J. Grim, Gregory Smith, Robert Ruby, and Alison Pond. *Spirit and Power: A Ten-Country Survey of Pentecostals.* Washington, DC: Pew Forum on Religion and Public Life, 2006.

Macchia, Frank D. *Spirituality and Social Liberation: The Message of the Blumhardts in the Light of Wuerttemberg Pietism.* Metuchen, NJ: The Scarecrow Press, 1993.

Mahler, Sarah J. "Engendering Transnational Migration: A Case Study of Salvadorans." *American Behavioral Scientist* 42, no. 2 (1998): 690–719.

Maier, Donna. "Nineteenth-Century Asante Medical Practices." *Comparative Studies in Society and History* 21, no. 1 (1979): 63–81.

Manuh, Takyiwaa. "Doing Gender Work in Ghana." In Cole, Manuh, and Miescher, *Africa After Gender?*, 125–49.

———. "Ghanaian Migrants in Toronto, Canada: Care of Kin and Gender Relations." *Ghana Studies* 6 (2003): 91–107.

———. "'This Place Is Not Ghana': Gender and Rights Discourse among Ghanaian Men and Women in Toronto." *Ghana Studies* 2 (1999): 77–95.

Marshall, Ruth. *Political Spiritualities: The Pentecostal Revolution in Nigeria.* Chicago: University of Chicago Press, 2009.

Marx, Karl. "Towards a Critique of Hegel's Philosophy of Right: Introduction." 1844. In *The Marx–Engels Reader*, edited by Robert C. Tucker, 53–65. New York: W. W. Norton, 1978.

Maxwell, David. *African Gifts of the Spirit: Pentecostalism and the Rise of a Zimbabwean Transnational Religious Movement.* Oxford: James Currey, 2006.

Maxwell, John, ed. *The Gold Coast Handbook, 1928.* London: Crown Agents for the Colonies, 1928.

Matory, J. Lorand. *Black Atlantic Religion: Tradition, Transnationalism, and Matriarchy in the Afro-Brazilian Candomblé.* Princeton: Princeton University Press, 2005.

Mazzucato, Valentina, Bart van den Boom, and N. N. N. Nsowah-Nuamah. "Origin and Destination of Remittances in Ghana." In *At Home in the World?; International Migration and Development in Contemporary Ghana and West Africa*, edited by Takyiwaa Manuh, 139–52. Accra: Sub-Saharan Publishers, 2005.

McAlister, Elizabeth. *Rara!: Vodou, Power, and Performance in Haiti and Its Diaspora.* Berkeley and Los Angeles: University of California Press, 2002.

McCaskie, T. C. "Akwantemfi—'in Mid Journey': An Asante Shrine Today, and Its Clients." *Journal of Religion in Africa* 38, no. 1 (2008): 57–80.

———. "Anti-Witchcraft Cults in Asante: An Essay in the Social History of an African People." *History in Africa* 8 (1981): 125–54.

———. *State and Society in Pre-Colonial Asante.* Cambridge: Cambridge University Press, 1995.

McElmurry, Beverly J., Karen Solheim, Rieko Kishi, Marcia A. Coffia, Wendy Woith, and Poolsuk Janepanish. "Ethical Concerns in Nurse Migration." *Journal of Professional Nursing* 22, no. 4 (2006): 226–35.

Meyer, Birgit. "'If You Are a Devil, You Are a Witch, and, If You Are a Witch Then You Are a Devil': The Integration of 'Pagan' Ideas into the Conceptual Universe of Ewe Christians in Southeastern Ghana." *Journal of Religion in Africa* 22, no. 2 (1992): 98–132.

———. "'Make a Complete Break with the Past': Memory and Postcolonial Modernity in Ghanaian Pentecostal Discourse." *Journal of Religion in Africa* 28, no. 3 (1998): 316–49.

———. *Translating the Devil: Religion and Modernity among the Ewe in Ghana.* Trenton, NJ: Africa World Press, 1999.
Meyerowitz, Eva L. R. *The Akan of Ghana: Their Ancient Beliefs.* London: Faber and Faber, 1958.
Middleton, John. "One Hundred and Fifty Years of Christianity in a Ghanaian Town." *Africa* 53, no. 3 (1983): 2–19.
Miescher, Stephan F. "Becoming ɔpinyin: Elders, Gender, and Masculinities in Ghana since the Nineteenth Century." In Cole, Manuh, and Miescher, *Africa After Gender?*, 253–69.
Mikell, Gwendolyn. *Cocoa and Chaos in Ghana.* Washington, DC: Howard University Press, 1989.
Miller, Jon. *Missionary Zeal and Institutional Control: Organizational Contradictions in the Basel Mission on the Gold Coast, 1828–1917.* Grand Rapids, MI: William B. Eerdmans, 2003.
Mohr, Adam. "Capitalism, Chaos, and Christian Healing: Faith Tabernacle Congregation in Southern Colonial Ghana, 1918–1926." *Journal of African History* 51, no. 1 (2011): 63–83.
———. "Out of Zion and Into Philadelphia and West Africa: Faith Tabernacle Congregation, 1897–1925." *Pneuma* 32, no. 1 (2009): 56–79.
———. "The Roaring Twenties: Faith Tabernacle Congregation, Emergent Pentecostalism, and Religious Revolution in Colonial West Africa, 1918–1929." Unpublished manuscript.
———. "School of Deliverance: Healing, Exorcism, and Male Spirit Possession in the Ghanaian Presbyterian Diaspora." In *Medicine, Mobility, and Power in Global Africa: Transnational Health and Healing*, edited by Hansjorg Dilger, Abdoulaye Kane, and Stacey Langwick, 241–70. Bloomington: Indiana University Press, 2013.
Morsey, Soheir A. "Spirit Possession in Egyptian Ethnomedicine: Origins, Comparisons, and Historical Specificity." In *Women's Medicine: The Zar-Bori Cult in Africa and Beyond*, edited by I. M. Lewis, Ahmed Al-Safi, and Sayyid Hurreiz, 189–208. Edinburgh: Edinburgh University Press, 1991.
Mullings, Leith. *Therapy, Ideology, and Social Change: Mental Healing in Urban Ghana.* Berkeley and Los Angeles: University of California Press, 1984.
Nancy, Jean-Luc. *Dis-Enclosure: The Deconstruction of Christianity.* New York: Fordham University Press, 2008.
Newell, Stephanie. *Literary Culture in Colonial Ghana: "How to Play the Game of Life."* Bloomington: Indiana University Press, 2002.
Obeng, Pashington. "Gendered Nationalism: Forms of Masculinity in Modern Asante of Ghana." In *Men and Masculinities in Modern Africa*, edited by Lisa A. Lindsay and Stephan F. Miescher, 192–208. Portsmouth, NH: Heinemann, 2003.

Obeng-Odoom, Franklin. "Land Reforms in Africa: Theory, Practice, and Outcome." *Habitat International* 36, no. 1 (2012): 161–70.
Okai, Samuel. *General History of Faith Tabernacle Congregation Ghana*. Accra: Faith Tabernacle Publishers, 2002.
Okoro, Johnson I. *Report on the International Presiding Elder of Faith Tabernacle Congregation: Pastor Kenneth W. Yeager's Third Visit to Africa (Nigeria and Ghana)*. Aba, Nigeria: Faith Tabernacle Congregation Printing Press, 2004.
Olupona, Jacob K., and Regina Gemignani. "Introduction." In Olupona and Gemignani, *African Immigrant Religions in America*, 1–26.
Olupona, Jacob K., and Regina Gemignani, eds. *African Immigrant Religions in America*. New York: New York University Press, 2007.
Omenyo, Cephas. "New Wine in an Old Wine Bottle?: Charismatic Healing in the Mainline Churches in Ghana." In Brown, *Global Pentecostal and Charismatic Healing*, 231–50.
———. *Pentecost Outside Pentecostalism: A Study of the Development of Charismatic Renewal in the Mainline Churches in Ghana*. Zoetermeer: Uitgeverij Boekencentrum, 2002.
Ong, Aihwa. "The Production of Possession: Spirits and the Multinational Corporation in Malaysia." *American Ethnologist* 15, no. 1 (1988): 28–42.
Onyinah, Opoku. "Akan Witchcraft and the Concept of Exorcism in the Church of Pentecost." PhD diss., University of Birmingham, 2002.
———. "The Man James McKeown." In *James McKeown Memorial Lectures*, 55–134. Accra: The Church of Pentecost, 2004.
Overa, Ragnhild. "When Men Do Women's Work: Structural Adjustment, Unemployment, and Changing Gender Relations in the Informal Economy of Accra, Ghana." *Journal of Modern African Studies* 45, no. 4 (2007): 539–63.
Oxfam. "Don't Copy Ghana's Health Insurance—Oxfam Warns Poor Countries." 2011.
Parker, John. "Witchcraft, Anti-Witchcraft, and Trans-Regional Ritual Innovation in Early Colonial Ghana: Sakrabundi and Aberewa, 1889–1910." *Journal of African History* 45, no. 3 (2004): 393–420.
Parsons, Talcott. "Introduction." In Weber, *Sociology of Religion*, xxix–lxxvii.
Patterson, K. David. "The Demographic Impact of the 1918–19 Influenza Pandemic in Sub-Saharan Africa: A Preliminary Assessment." In *African Historical Demography II*, edited by Christopher Fyfe and David McMaster, 403–31. Edinburgh: University of Edinburgh Press, 1981.
———. "The Influenza Epidemic of 1918–19 in the Gold Coast." *Journal of African History* 24, no. 4 (1983): 485–502.
Patterson, K. David, and Gerald W. Hartwig. "The Disease Factor: An Introductory Overview." In *Disease in African History*, edited by Gerald

W. Hartwig and K. David Patterson, 3–24. Durham: Duke University Press, 1978.

Patterson, K. David, and Gerald F. Pyle. "The Geography and Mortality of the 1918 Influenza Pandemic." *Bulletin of the History of Medicine* 65, no. 1 (1991): 4–21.

Peel, John D. Y. *Aladura: A Religious Movement among the Yoruba.* London: Oxford, 1968.

———. *Religious Encounters and the Making of the Yoruba.* Bloomington: Indiana University Press, 2003.

Peil, Margaret. "Ghanaians Abroad." *African Affairs* 94, no. 376 (1995): 345–67.

Pelrez y Mena, Andres. *Speaking with the Dead; Development of Afro-Latin Religion among Puerto Ricans in the United States.* New York: AMS Press, 1991.

Pessar, Patricia R. "Engendering Migration Studies: The Case of New Immigrants in the United States." *American Behavioral Scientist* 42, no. 4 (1999): 577–600.

———. "On the Homefront and in the Workplace: Integrating Immigrant Women into Feminist Discourse." *Anthropological Quarterly* 68, no. 1 (1995): 37–47.

Pew Research Center. *Global Christianity: A Report on the Size and Distribution of the World's Christian Population.* Washington, DC: Pew Research Center's Forum on Religion and Public Life, 2011.

Pittman, Patricia, Linda H. Aiken, and James Buchan. "International Migration of Nurses: Introduction." *Health Service Research* 42, no. 3, part 2 (2007): 1275–80.

Presbyterian Church of Ghana. "Appendix F: Report on Prayer Groups and Sects." *Minutes of the 37th Synod*, 41–54. Akropong: Presbyterian Church of Ghana, 1966.

Presbyterian Church of the USA. "Presbyterian 101." 2007. www.pcusa.org/101-sacrament.html.

Ramsey, Kate. *The Spirits and the Law: Vodou and Power in Haiti.* Chicago: University of Chicago Press, 2011.

Ranger, Terence. "Godly Medicine: The Ambiguities of Medical Mission in Southeastern Tanzania, 1900–1945." In Feierman and Janzen, *Social Basis of Health and Healing in Africa*, 256–84.

Rattray, Robert S. *Ashanti.* Oxford: Clarendon Press, 1923.

———. *Religion and Art in Ashanti.* Oxford: Clarendon Press, 1927.

Reindorf, Carl C. *History of the Gold Coast and Asante.* Basel: Basel Mission Press, 1985.

Richman, Karen. *Migration and Vodou.* Gainesville: University of Florida Press, 2005.

Riss, Richard M. "Latter Rain Movement." In *The New International Dictionary of Pentecostal Charismatic Movements*, edited by Stanley M. Burgess, 830–33. Grand Rapids, MI: Zondervan, 2003.

Ross, S. J., D. Polsky, and J. Sochalski. "Nursing Shortages and International Nurse Migration." *International Nursing Review* 52, no. 4 (2005): 253–62.

Saler, Michael. "Modernity and Enchantment: A Historiographic Review." *American Historical Review* (June 2006): 692–716.

Sassen, Saskia. "Global Migration and Economic Need." Paper given to the Penn Program on Democracy, Citizenship, and Constitutionalism, University of Pennsylvania, October 25, 2007.

Schmidt, Bettina E. "Possessed Women in the African Diaspora: Gender Difference in Spirit Possession Rituals." In Schmidt and Huskinson, *Spirit Possession and Trance*, 97–116.

Schmidt, Bettina E., and Lucy Huskinson, eds. *Spirit Possession and Trance: New Interdisciplinary Perspectives*. London: Continuum International, 2010.

Schrock, Douglas, and Michael Schwalbe. "Men, Masculinity, and Manhood Acts." *Annual Review of Sociology* 35 (2009): 277–95.

Schwalbe, Michael L. "Identity Stakes, Manhood Acts, and the Dynamics of Accountability." In *Studies in Symbolic Interaction*, edited by Norman K. Denzin, 65–81. New York: Elsevier, 2005.

Scripture Union Prayer Warrior Ministry. *Twentieth Anniversary, Healing and Deliverance Workshop, 2004*. Accra: SonLife Printing Press, 2004.

Shankar, Shobana. "Medical Missionaries and Modernizing Emirs in Colonial Hausaland: Leprosy Control and the Native Authority in the 1930s." *Journal of African History* 48, no. 1 (2007): 45–68.

Shipton, Parker. *Mortgaging the Ancestors: Ideologies of Attachment in Africa*. New Haven: Yale University Press, 2009.

Silverstein, Michael. "Metaforces of Power in Traditional Oratory." Lecture, Department of Anthropology, Yale University, 1981.

Smith, Noel. *The Presbyterian Church of Ghana: 1835–1960*. Accra: Ghana University Press, 1966.

Soothill, Jane E. *Gender, Social Change, and Spiritual Power: Charismatic Christianity in Ghana*. Leiden: Brill, 2007.

Southall, Roger J. "Cadbury on the Gold Coast, 1907–1938: The Dilemma of the 'Model Firm' in a Colonial Economy." PhD diss., University of Birmingham, 1975.

Sundkler, Bengt, and Christopher Steed. *A History of the Church in Africa*. Cambridge: Cambridge University Press, 2000.

Sylvester, Nigel. *God's Word in a Young World: The Story of Scripture Union*. London: Scripture Union, 1984.

Ter Haar, Gerrie. "Ghanaian Witchcraft Belief: A View from the Netherlands." In *Imagining Evil: Witchcraft Beliefs and Accusations in*

Contemporary Africa, edited by Gerrie ter Haar, 93–112. Trenton, NJ: Africa World Press, 2007.

———. *Halfway to Paradise: African Christians in Europe.* Cardiff: Cardiff Academic Press, 1998.

———. *How God Became African: African Spirituality and Western Secular Thought.* Philadelphia: University of Pennsylvania Press, 2009.

Terrazas, Aaron. "African Immigrants in the United States." *Migration Information Source (MPI).* 2009.

Tiilikainen, Marja. "Somali *Saar* in the Era of Social and Religious Change." In Schmidt and Huskinson, *Spirit Possession and Trance,* 117–33.

Tsikata, Dzodzi. "Women's Organizing in Ghana since the 1990s: From Individual Organizations to Three Coalitions." *Development* 52, no. 2 (2009): 185–92.

Turner, Bryan S. *Religion and Modern Society: Citizenship, Secularisation, and the State.* Cambridge: Cambridge University Press, 2011.

Turner, Harold W. *History of an African Independent Church: The Church of the Lord (Aladura).* Oxford: Clarendon Press, 1967.

Twum-Baah, Kwaku. "Volume and Characteristics of International Ghanaian Migration." In *At Home in the World: International Migration and Development in Contemporary Ghana and West Africa,* edited by Takyiwaa Manuh, 55–77. Accra: Sub-Saharan Publishers, 2005.

Ubink, Janine M. "Negotiated or Negated?: The Rhetoric and Reality of Customary Tenure in an Ashanti Village in Ghana." *Africa* 78, no. 2 (2008): 264–87.

Van Binsbergen, Wim. "Witchcraft in Modern Africa as Virtualized Boundary Condition of the Kinship Order." In *Witchcraft Dialogues: Anthropological and Philosophical Exchanges,* edited by George Clement Bond and Diane M. Ciekawy, 212–63. Athens: Ohio University Press, 2001.

Van Dijk, Rijk. "From Camp to Encompassment: Discourses of Transubjectivity in the Ghanaian Pentecostal Diaspora." *Journal of Religion in Africa* 27, no. 2 (1997): 135–59.

Vaughan, Megan. *Curing Their Ills: Colonial Power and African Illness.* Stanford: Stanford University Press, 1991.

Walls, Andrew F. "Eusebius Tries Again: Reconceiving the Study of Christian History." *International Bulletin of Missionary Research* 24, no. 3 (2000): 105–11.

Weber, Marianne. *Max Weber: A Biography.* Edited and Translated by Harry Zhon. New York: Wiley, 1975.

Weber, Max. *From Max Weber: Essays in Sociology.* Edited and Translated by H. H. Gerth and C. Wright Mills. New York: Oxford University Press, 1946.

———. *The Protestant Ethic and the Spirit of Capitalism.* 1904–5. Edited and translated by Talcott Parsons. New York: Charles Scribner's Sons, 1976.

———. *The Sociology of Religion*. 1922. Translated by Ephraim Fischoff. Boston: Beacon Press, 1993.

Westerlund, David. *African Indigenous Religion and Disease Causation: From Spiritual Beings to Living Humans*. Leiden: Brill, 2006.

Wilks, Ivor. *Forests of Gold: Essays on the Akan and the Kingdom of Asante*. Athens: Ohio University Press, 1993.

Williamson, Sidney George. *Akan Religion and the Christian Faith*. Accra: Ghana University Press, 1965.

Wirtz, Kristina. *Ritual, Discourse, and Community in Cuban Santeria: Speaking the Sacred Word*. Gainesville: University of Florida Press, 2005.

Women's Law Project. "Violence Against Women: Impact on Public Policy." 2011. www.womenslawproject.org/NewPages/wkVAW_policy.html.

Wong, Madeleine. "The Gendered Politics of Remittances in Ghanaian Transnational Families." *Economic Geography* 82, no. 4 (2005): 355–81.

Wrigley-Asante, Charlotte. "Men Are Poor but Women Are Poorer: Gendered Poverty and Survival Strategies in the Dangme West District of Ghana." *Norwegian Journal of Geography* 62, no. 3 (2008): 161–70.

Wyllie, Robert W. *Spirit Seekers: New Religious Movements in Southern Ghana*. Missoula: Scholars Press, 1980.

Yeboah, Muriel Adjubi. "Urban Poverty, Livelihood, and Gender: Perceptions and Experiences of Porters in Accra, Ghana." *Africa Today* 56, no. 3 (2010): 43–60.

Zehring, Fred. "Our Church History (Faith Tabernacle Congregation)." Term paper, Faith Tabernacle Secondary School, Harrisburg, PA, 1974.

Zentgraf, Kristine. "Immigration and Women's Empowerment: Salvadorans in Los Angeles." *Gender and Society* 16, no. 5 (1998): 625–46.

Zimmerman, Andrew. *Alabama in Africa: Booker T. Washington, the German Empire, and the Globalization of the New South*. Princeton: Princeton University Press, 2010.

Zuendel, Freidrich. *One Man's Battle with Darkness*. 1880. Farmington, PA: The Plough Publishing House, 2000.

Index

Abboah-Offei, Catechist Ebenezer, 1–2, 15–16, 30, 83, 85n11, 94, 94n47, 95–96, 96nn53–54, 97, 97n56–58, 98–101, 101nn60–61, 102, 106, 106n64, 107–9, 109n72, 120–23, 123nn33–35, 124, 124n37, 125, 127–32, 134, 136–37, 145, 148–65, 165n45, 166–67, 176, 183, 183n57, 186–87, 187n60, 195–96, 206, 206n35

Aboagye, Rev. Margaret Asabea, 179–82, 182n56, 183–84, 190

Accra, 13, 22, 24, 61–62, 78, 90, 95, 106, 114, 127, 138, 142, 160

aduru, 34, 35n49, 197

aduruyefo, 34, 35n49, 36, 43, 46, 186, 186n59, 187, 197, 197n4, 198, 206

Afwireng, Catechist N., 41, 43–46

Akan healer, 14, 35n49, 38n63, 39–40, 40n70, 42, 42n82, 44, 46–53, 61, 154, 164, 189, 194, 198, 200

Akan therapeutic practitioner, 43, 164. *See also* Akan therapeutics

Akan therapeutics, 31, 31n36, 40, 40n70, 52, 154

Akropong, 13, 22–23, 29–30, 44–45, 49, 53, 56, 60, 76, 83, 94, 94n47, 95–97, 99, 104, 121–22, 124, 138, 149, 163, 168, 176, 183, 195

Akwapim, 94, 94n47, 116, 145, 160

American Exceptionalism, 202

Anim, Peter, 71, 71n88, 72, 75, 75n114, 76, 80

antiwitchcraft shrine/cult, 59, 59n31, 59n33

Apostolic Church, 73, 75–78, 80, 87, 118n23, 143

Asamankese, 56, 70n85, 71–72, 75, 87, 94, 95, 121

Asante, 23, 23n9, 32n38, 55–56, 56n14, 61–62, 62n46, 70, 75, 81, 90, 115–16, 116n20, 144

Asante, Rev. David, 94

Asare, Samuel, 2, 97, 105, 125–26, 129–30, 130n53, 131, 133, 159, 184, 187

Atiemo, Rev. Samuel, 30, 79, 121, 122, 122n32, 123n33, 124, 126, 129, 149, 206n35

Atta-Fynn, Rev. Rosamund, 79, 147, 153, 158–59, 166–67, 181–85, 187, 190

bacteriological revolution, 6, 46, 47, 52, 194, 199, 203n27

Bad Boll, 9n26, 24, 27–30, 40, 43, 46–47, 194

Basel Mission, 9, 13, 21–23, 24n12, 25n16, 28–30, 33, 38, 39–40, 40n68, 44, 46, 48, 49n103, 51–55, 55n13, 56–58, 58n30, 59–61, 65–66, 68, 70–71, 81, 83, 94, 194, 196–200

Basel missionaries, 13–14, 14n38, 21–24, 28–30, 39–41, 43, 47–51, 55–56, 60, 62, 66, 66n68, 81, 194

Bible Study and Prayer Group (BSPG), 88–89, 89n25, 90, 92, 94, 96, 96n53, 97–98, 108, 120, 132, 134, 137, 139, 141, 147, 165, 167, 195, 206nn35–36

biomedicine, 14, 16, 42, 46–48, 50, 52, 75, 85, 87, 87n19, 88, 98, 121, 128–29, 154–55, 165, 194, 196–201, 207

Bismarck, Otto von, 202, 203n27

Blumhardt, Christian Gottlieb, 24

Blumhardt, Johann Christoph, 9, 25, 25n16, 26–28, 47, 52, 64, 194

Calvin, John, 4

Calvinism, 2, 4, 6, 8–10, 12, 107, 109, 134, 167, 169, 206n35, 207
capitalism, 2, 4, 5, 5n9, 6, 8, 12, 14, 17, 85, 173–75, 193–94, 196–97, 201, 203, 203n27, 205, 207
"cash-and-carry" medical payment system, 84, 200
Challenge Enterprises bookstore, 91–96, 108–9, 109n72
Christ Apostolic Church, 76, 87, 87n19, 94, 140n6
Christaller, J. G., 146
Christian healing, 9, 9n26, 15, 24–25, 29–30, 46–48, 85, 87, 123, 155, 170–71, 184, 194, 196. *See also* Christian therapeutics
Christian therapeutics, 14, 175n28. *See also* Christian healing
Christ Presbyterian Church, 96, 96n53
Church of Pentecost, 13, 16, 77–81, 86–88, 94, 96–97, 101, 118–20, 120n27, 143–44, 144n19, 189, 195
Clark, Ambrose, 64n51, 67–69, 69n83, 71–73, 73n103, 74
cocoa, 10, 14, 15, 28, 53–54, 54n5–6, 55, 55n15, 56, 56n14, 57, 57n22, 58–63, 70–71, 76, 81, 84, 86, 94, 114–16, 121, 134, 137, 172–74, 194
communal land tenure, 204. *See also* customary land tenure; land tenure
communion, 163–65
Conference of Ghanaian Presbyterian Churches, North America, 117, 121n30, 124, 136, 145, 145n23, 148
customary land tenure, 202, 205, 207. *See also* communal land tenure; land tenure
customary law, 204–5

deliverance: counselor, 103–4; questionnaire, 101–3, 129, 132, 149, 153, 154n32, 162, 186; workshop (Ghana), 91–92, 96n54, 100–101, 108; workshop (New York), 13, 15, 15n39, 113, 120, 121n30, 122–25, 127–28, 131–33, 135–36, 146, 153, 155, 158–59, 166–67, 178, 199

demonic possession, 25, 64, 119, 128, 131, 155, 162, 164. *See also* spirit possession
demons, 25, 59n33, 131
devil, 24, 30, 39, 64, 119, 130–31, 133, 155
diagnose, 16, 101, 103, 152, 186
diagnosis, 26, 46, 103–5, 132, 155–56, 161
disenchantment, 2, 5, 6, 12, 21, 193, 196, 202–3, 203n27, 208
Dittus, Gottlieben, 25–26
divine healing, 9, 9n25–26, 15, 30, 62–65, 65n65, 66–67, 69, 71, 83n2, 128n51, 151, 199, 208
doctor(s), 197n4; biomedical, 25–26, 47, 85, 98, 151–52, 154, 158, 186, 200; African or traditional or native, 42–44, 102, 154n32; Basel Mission or missionary, 46, 48–49, 198, 198n8. *See also* physician(s)
Dowie, John Alexander, 9n25, 63, 65n65

Eastern Region, 55–56, 76, 79, 81, 83, 86–87, 116, 144
enchanted, 2, 6, 8–15, 21, 52–53, 59, 63, 72, 79, 80–83, 85–86, 94, 107–9, 113, 116–17, 123, 134–36, 167–69, 193–96, 199, 201, 206, 206n35, 208
enchantment, 6n14, 7–10, 12, 15, 17, 109, 113–14, 193–96, 201, 204, 207
Evans-Pritchard, E. E., 35n49, 186n59

Faith Home, 68, 68n75
Faith Tabernacle Congregation, 13, 15, 30, 59, 62–66, 66n68, 67–73, 73n103, 74–76, 80–81, 85–87, 98, 121, 137, 194, 198
"female senior masculinity," 189–91
First-Century Gospel Church, 73–75
Fisch, Dr. Rudolf, 46–49, 200
Frimpong-Manso, Dr. Rev. Yaw, 91, 107, 148

gender, 16, 52, 71, 134, 171, 171n9, 172n10, 175n27, 180, 183n57, 184, 189, 191–92, 196; egalitarianism,

170–71, 174, 176, 179–80, 181, 181n50
Grace Presbyterian Church, 15, 93, 97–107, 123, 126, 149, 151, 154, 158, 169, 176, 183, 183n57, 184
Grace Team, 96n53, 98–103, 106, 151. *See also* prayer team
healing home, 9n26, 24, 27, 29; divine, 27, 68, 68n75. *See also* Faith Home
health insurance, 85, 201, 207
HIV, 160–61, 163

ideal types, 6–8, 208
influenza pandemic of 1918–19, 14, 30, 60–62, 66–68, 81, 194

Joyful Way Incorporated, 95, 121

Kant, Immanuel, 6, 208
Kumasi, 23, 56, 68, 78, 87, 91, 95, 124, 198
Kumi, T. A., 88, 97
Kurhaus, 24, 27, 194

land tenure, 17, 196, 204. *See also* communal land tenure; customary land tenure
labor migrants, 1n2, 2, 15, 17, 116, 134, 137, 190, 195, 204–5
Latter Rain, 77–78, 119

malaria, 22, 41–43, 49, 49n105, 199
matrilineage, 57, 58, 121, 138, 173, 175, 205
McKeown, James, 75–79, 88
medical missionaries, 47–49, 197. *See also* medical missions
medical missions, 147, 199. *See also* medical missionaries
Methodist Church, 93, 101, 127, 137–38, 143, 160
Moettlingen, 25–28

Neo-Pentecostal, 101, 108–9, 122–23, 127, 137, 139, 191
Nigeria, 60, 65, 68–69, 77, 92, 108, 115, 140, 191, 195

Nti, Kwadjo, 68, 68n79, 198
nurse(s), 49, 85, 98, 99n59, 154, 177, 177n37, 178, 178n40, 198. *See also* nursing
nursing, 126, 177, 177n37, 178, 178n43. *See also* nurse(s)

Ofosu-Donkoh, Dr. Rev. Kobina, 121n30, 137–49, 152, 158–59, 163–68, 182, 184, 188, 207
Okyere, Edward, 30, 90–92, 97, 107, 206n35

pastoral care, 1, 8–9, 88, 166
Pentecostalism, 9, 75–76, 81, 137, 165
physician(s), 199; biomedical, 16, 26–27, 47, 85, 85n11, 106, 151–52, 155, 199–201; colonial, 42, 48, 61, 197–98; native, 36n58, 38n63. *See also* doctor(s)
prayer camp, 96, 97, 120, 120n28, 126, 189
prayer team, 2, 89n25, 91, 100–101, 107n65, 120, 122–25, 129, 132, 134–37, 145–48, 152–53, 155–63, 165–67, 167n48, 168, 181, 183–84, 187, 195–96. *See also* Grace Team
Presbyterian Church of Canada, 117–18, 123
Presbyterian Church of Ghana, 1, 9–10, 12–13, 15–16, 24, 30, 53, 60, 79–81, 83–93, 97–98, 107–9, 117, 121, 123, 126, 132, 134, 139, 141, 142, 146–48, 164–65, 167, 169, 182, 187, 189, 195–96, 201
Presbyterian Church of the USA, 11n33, 117–18, 123, 141–44, 147, 164
Protestant Ethnic and the Spirit of Capitalism, 2, 5, 12, 201

Quashie, Tetteh, 54
quinine, 49, 49n105, 50, 199

Radio ELWA, 91–92
Riis, Andreas, 22, 41–42, 42n82, 43, 49, 96

Sam, Joel Sackey, 66–67, 71, 74

sanitary segregation, 199
Scripture Union, 30, 89–92, 94–95, 97, 101–2, 107–8, 121, 124, 126, 137–39, 165
social welfare, 196, 201, 203
socio-spiritual afflicition, 15, 17, 205, 206, 207
spirit possession, 25n16, 64n53, 100, 113–14, 155–56, 163, 169–71, 171n9, 174, 175n27, 176, 181, 185, 190–91; female, 175; male, 16, 184, 187, 190, 192, 196. *See also* demonic possession
state, 8, 17, 174, 179–80, 201, 204–7
Stony Point Center for Education and Mission, 122–25, 129, 142, 153
Sudan Interior Mission, 89, 91–92

Tabiri, Owusu, 96, 96n54, 120, 120n27
Tetteh Quashie Memorial Hospital, 99, 106
Transnational, 113–15, 121, 132, 134, 196–98, 205–8
Trudel, Dorothea, 27–29

United Ghanaian Community Church (UGCC), 15–16, 118, 136–37, 139, 142–49, 153–54, 158–68, 181, 183–84, 186–87, 195–96, 204

Weber, Max, 2, 4–9, 12, 12n37, 193, 196, 201, 203, 208
welfare spending, 17, 201–4, 207
welfare state, 202–4
Western Europe, 4–5, 12, 116, 202
Wigglesworth, Smith, 128, 199
Wilhide, Jacob Thomas, 63–64
Winneba, 63, 66–68, 72, 74–75, 137
witch, 34, 36n54, 130n55. *See also* witchcraft
witchcraft, 14–15, 53, 58–59, 81, 101, 103, 106–7, 113–14, 130, 133–34, 137, 148, 154, 194, 205. *See also* witch
worship team, 147–48, 159
Württemberg Pietism, 21, 24, 27–28, 39, 47–48

Yeboah, M. K., 79–80, 97, 97n57